5

Nanotubes and Related Nanostructures–2014

MATERIALS RESEARCH SOCIETY
SYMPOSIUM PROCEEDINGS VOLUME 1700

Nanotubes and Related Nanostructures–2014

Symposium held April 21-25, 2014, San Francisco, California, USA

EDITORS

Don Futaba

National Institute of Advanced Industrial
Science and Technology
Ibaraki, Japan

Yoke Khin Yap

Michigan Technological University
Houghton, Michigan, USA

Materials Research Society
Warrendale, Pennsylvania

CAMBRIDGE UNIVERSITY PRESS
Cambridge, New York, Melbourne, Madrid, Cape Town,
Singapore, São Paulo, Delhi, Mexico City

Cambridge University Press
32 Avenue of the Americas, New York, NY 10013-2473, USA

www.cambridge.org
Information on this title: www.cambridge.org/9781605116778

Materials Research Society
506 Keystone Drive, Warrendale, PA 15086
http://www.mrs.org

© Materials Research Society 2014

First published 2014

CODEN: MRSPDH

ISBN: 978-1-60511-677-8 Hardback

CONTENTS

SYNTHESIS AND STRUCTURE

ELECTRICAL INVESTIGATION

*Invited Paper

*Invited Paper

vii

PREFACE

Carbon nanotubes and related nanostructures, including nanosheets have attracted tremendous attention for their unique structures and intriguing properties. These nanomaterials have been widely investigated—from theory, synthesis, and characterization to applications in electronic devices, energy generation and storage, biological and chemical sensors, etc. In addition, non-carbon nanostructures such as nanotubes and nanosheets of boron nitride (BN) have gained increasing interest.

To facilitate scientific interaction among students and researchers on the latest advancements in this area, Symposium MM – Nanotubes and Related Nanostructures, was organized and held on Apr. 21–25 at the 2014 MRS Spring Meeting in San Francisco, California. The symposium organizers were Don Futaba (National Institute of Advanced Industrial Science and Technology), Annick Loiseau (Laboratoire d'Etude des Microstructures), Yoke Khin Yap (Michigan Technological University), and Ming Zheng (National Institute of Standards and Technology).

This proceedings volume consists of peer-reviewed papers presented in the symposium, including invited and contributed presentations. These papers represent a snapshot of topics discussed in both theoretical and experimental aspects. We hope this publication will contribute toward productive research in the area of nanotubes and related nanostructures.

Don N. Futaba
Yoke Khin Yap

September 2014

Acknowledgments

The papers published in this volume result from the MRS Spring 2014 symposium MM. We extend our gratitude to all the oral and poster presenters of the symposium who contributed to this volume. We also thank the reviewers of these manuscripts, who provided valuable feedback to the editors and authors. The organizers of Symposium MM thank Hummingbird Scientific, the Multi-Scale Technologies Institute (MuSTI) at Michigan Technological University for their financial support.

Y.K. Yap acknowledges the National Science Foundation (DMR 1261910) for supporting his outreach efforts to promote the interest of the younger generation in science, technology, engineering, and mathematics (STEM) education.

MATERIALS RESEARCH SOCIETY SYMPOSIUM PROCEEDINGS

MATERIALS RESEARCH SOCIETY SYMPOSIUM PROCEEDINGS

Prior Materials Research Symposium Proceedings available by contacting Materials Research Society

Synthesis and Structure

Mater. Res. Soc. Symp. Proc. Vol. 1700 © 2014 Materials Research Society
DOI: 10.1557/opl.2014.604

Support Vector Machine Classification of Single Walled Carbon Nanotube Growth Parameters

N. Westing[1], J. Clark[1] and D. Hooper[2], P. Nikolaev[2], B. Maruyama[3]

[1]Dept. of Electrical and Computer Engineering, Air Force Institute of Technology, 2950 Hobson Way Wright-Patterson AFB, OH 45433
[2]UES, Inc., 4401 Dayton-Xenia Rd. Dayton, Ohio 45432
[3]Air Force Research Laboratory, Materials and Manufacturing Directorate, RXAS, Wright-Patterson AFB, OH 45433

ABSTRACT

Selective single-walled carbon nanotube (SWNT) growth is a challenging problem, limiting their use in a wide variety of applications. Significant degrees of freedom in these experiments may lead to synthesis of multi-walled carbon nanotubes (MWNTs), which are less preferred. Thus, a method for constraining the synthesis results to only SWNTs is desired. A machine learning based approach for selectively growing SWNTs using a laser-induced chemical vapor deposition growth system is introduced. This approach models the complex relationships between the associated synthesis parameters to predict SWNT growth. The parameters under consideration include argon, ethylene, hydrogen and carbon dioxide partial pressures, growth temperature, and water vapor concentration. The catalyst consists of 10 nm of alumina and 1 nm of nickel deposited onto 10 μm diameter silicon pillars with a height of 10 μm. Determination of SWNT growth is performed through in-situ Raman spectroscopy using a 532 nm excitation laser. A total of 121 experiments are used to train a SWNT vs. MWNT classifier with a resulting model accuracy of 94.21%. The classifier model is applied to a range of simulated inputs, and the subset of these inputs that meet a >90% probability of SWNT growth are investigated further. The simulated inputs consist of 531,201,645 unique growth parameter combinations spanning the entire parameter space. A reduced dataset of 449,117 growth parameter combinations define 90% probability of SWNT growth according to the model. Randomly selected input parameters from this reduced dataset were tested experimentally, resulting in SWNT growth for all performed experiments validating the classifier model. This approach maps input growth conditions to SWNT growth selectivity using a limited set of experimental data and allows for further investigation into SWNT growth rates and chiral dependencies.

INTRODUCTION

The electronic properties of single walled carbon nanotubes (SWNTs) have been extensively studied and shown to depend greatly on their diameter and chirality [1]. Several methods exist for chiral-selective growth of a single walled carbon nanotube (SWNT) [2-4]. Recently, an approach has been found for influencing the chirality of a SWNT by controlling the growth rate [5]. Growth parameters for SWNTs using a CVD growth system are well established and much of the growth characteristics are well known [6]. Adjustment of these growth parameters results

The views expressed in this paper are those of the authors and do not reflect the official policy or position of the United States Air Force, Department of Defense, or the United States Government. The authors thank the Air Force Research Laboratory/Materials and Manufacturing Directorate, WPAFB for the use of their in-situ Raman spectroscopy CNT growth system and ex-situ Raman measurements.

in varying growth rates and catalyst lifetimes [7]. Adjusting these parameters typically results in SWNT growth however multi-walled carbon nanotube (MWNT) growth will also occur if incorrect parameters are used [7]. Complete control of SWNT chirality requires finding growth parameter regions with the highest likelihood of SWNT growth. An exhaustive experimental search across the region is intractable due to the large number of growth parameters; however an attractive solution for finding these parameter space boundaries is accomplished using a Support Vector Machine (SVM) classifier (explained below). The adjustable growth parameters consist of: ethylene, hydrogen, argon, and carbon dioxide flow rates, chamber pressure, growth temperature and water vapor concentration. The growth parameter regions meeting a >90% probability of SWNT growth are examined further. The role of the catalyst is not directly explored in this work; however, training an SVM classifier for each new catalyst will provide insight into the roles different catalysts play in the growth process. For the catalyst studied in this work, the SVM classifier is able to determine unique growth regions between each of the growth parameters.

THEORY

The SVM classifier creates a separating hyperplane between the MWNT and SWNT datasets [8]. The hyperplane location is at the maximum distance from the nearest point in each class based on the large margin classifier concept. A margin is placed around the hyperplane, where the points lying on this margin are considered support vectors [8].

In nonlinearly separable datasets a kernel function is used to project the dataset to a higher dimensional feature space, creating a feature space where the hyperplane can separate the classes [8]. The kernel function used in this work is a radial basis function (RBF), as described by Eq. 1, where σ is the variance, x_i is an input growth parameter vector, x_j is another input growth parameter vector:

$$K(x_i, x_j) = \exp\left(-\frac{\|x_i - x_j\|^2}{2\sigma^2}\right). \tag{1}$$

The SVM model predicts SWNT or MWNT growth with a given probability for a set of growth parameters. Growth parameters near the SVM hyperplane are more likely to produce a combination of SWNTs and MWNTs. The probability of only SWNT growth increases as the distance from the SVM hyperplane is increased. John Platt's sequential minimization optimization algorithm (SMO) is used within the Weka [9,10] software package to train the classifier and obtain proper probabilities for each class.

The laser-induced CVD growth system, Adaptive Rapid Experimentation and in-situ Spectroscopy (ARES[1]), depicted in Fig. 1, is well suited for automated growth data collection. Samples consist of 10 μm diameter pillars of silicon placed 40 μm apart. The catalyst material consists of 10 nm of alumina and 1 nm of nickel. A 532nm excitation laser is used for silicon pillar heating and Raman spectroscopy. The laser is focused on each pillar using a 50X objective lens resulting in a laser spot size of 7 μm. Research grade argon, hydrogen, ethylene and carbon dioxide are pumped into the growth chamber at varying flow rates and the water vapor content is monitored. The growth parameters do not include laser power due to slight variations in pillar

[1]Patent Pending

height or laser focus resulting in different experimentation temperatures for a constant laser power. The resulting temperature is recorded and is calculated as follows [11]:

$$\Delta\omega(T) = C\left[1 + \frac{2}{e^{\frac{\hbar\omega_0}{2k_BT}} - 1}\right] + D\left[1 + \frac{3}{e^{\frac{\hbar\omega_0}{3k_BT}} - 1} + \frac{3}{\left(e^{\frac{\hbar\omega_0}{3k_BT}} - 1\right)^2}\right], \qquad (2)$$

where C and D are anharmonic constants, $\hbar$ is the Planck constant, T is the temperature, ω_0 is the silicon peak frequency, and k_B is the Boltzmann constant.

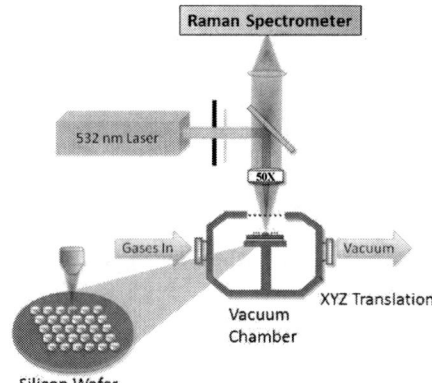

Figure 1. ARES[1] growth system setup

DISCUSSION

To classify SWNT growth versus MWNT growth, a wide range of the growth parameter space is explored through direct experimentation. These initial 121 growth experiments provide the training data for the SVM classifier. The growth experiments resulting in SWNT or MWNT growth are not linearly separable and require a kernel function to linearly separate the data. The kernel function shown in Eq. 1 requires a variance (σ) value. Very small σ values result in a dataset specific classifier with limited generalizability and too large of a σ value will classify poorly.

Figure 2 shows the number of support vectors and model accuracy as σ is adjusted from 0.1 to 100. A σ value of 6.31 is selected due to its ability to maximize accuracy, 94.21%, and area under the receiver operating characteristic (ROC) curve, 0.972, in relation to its neighbors while maintaining the fewest support vectors, 32.

A simulated dataset is created based on every combination of the growth parameters shown in Table I resulting in 531,201,645 unique growth parameter combinations. The partial pressure

---

[1] Patent Pending

of each gas is used for training and testing, therefore, the summation of these partial pressures must equal the total pressure used in each growth experiment.

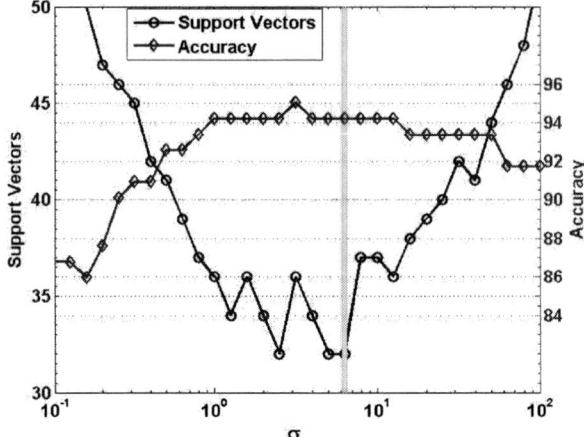

Figure 2. The σ value used in the SVM classifier is highlighted by the vertical line in the plot resulting in a model accuracy of 94.21% using 32 support vectors.

Table I. Simulated Range and Resolution

	Range	Resolution
C_2H_4 **(Torr)**	1 - 40	1
H_2 **(Torr)**	1 - 40	1
Ar/CO_2 **(Torr)**	1 - 40	1
Temperature(C)	300 - 1200	20
Water (ppm)	1 - 200	1

From the initial 531,201,645 simulated experiment set, 449,117 simulated experiments are predicted to produce SWNT growth with greater than a 90% probability, generating a reduced optimal dataset. A histogram of each feature in this reduced dataset is plotted to determine trends in optimal growth regions as shown in Fig. 3. It should also be noted that the other growth parameters are not arbitrary when considering Fig. 3. Each gas partial pressure has unique relationships with one another, temperature and water vapor concentration making visualizations of these hyper dimensional relationships impossible. The optimal temperature for MWNT growth shown in Fig. 3b is not expected to produce high quality tubes and is instead a result of experimental bias in the training data.

The relationship between ethylene and hydrogen partial pressures, and temperature was analyzed using the SVM classifier by holding the water concentration constant at 25 ppm and argon/carbon dioxide partial pressure constant at 4 Torr. Figure 4a is a contour plot of the probability of selective SWNT growth (i.e., no MWNT growth) as a function of ethylene and

---

[1]Patent Pending

hydrogen partial pressures, and temperature. Figure 4a illustrates a region within the multi-parametric process space which maximizes the predicted probability of selective SWNT growth. In Figure 4b the results are re-plotted such that the optimal temperature (i.e., that with the highest probability of selective SWNT growth) is shown as a function of hydrogen and ethylene partial pressures.

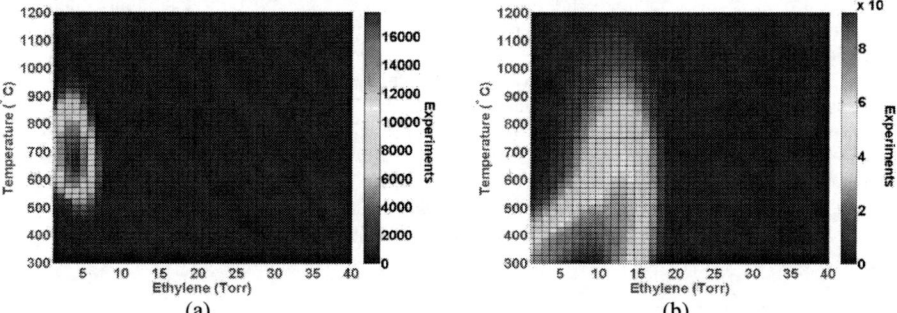

(a)   (b)

Figure 3. (a) Frequency of optimal growth temperature and ethylene partial pressure for 90% confident SWNT growth experiments. (b) Frequency of optimal experiments for MWNT growth experiments. All other growth parameters are also changing to form these plots.

In order to validate the SVM classifier 20 experimental conditions were randomly chosen from the reduced optimal dataset as inputs for experimentation. All 20 experiments resulted in selective SWNT growth, thus validating the efficacy of our SVM classifier. Future work will include an SVM sensitivity analysis and other feature selection methods to verify which parameters have the strongest effect on growth.

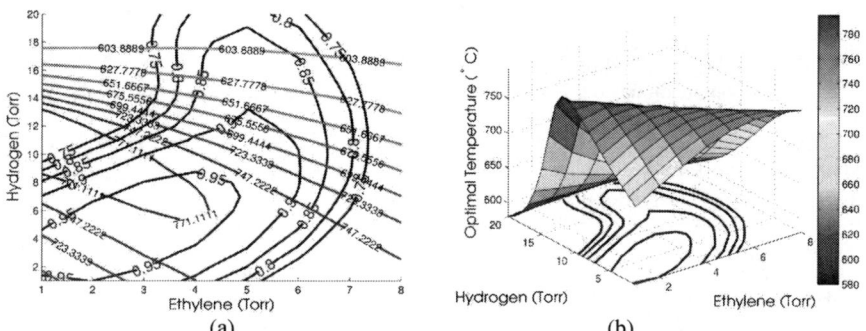

(a)   (b)

Figure 4. (a) Optimal SWNT growth temperature contour overlaid on the probability of SWNT growth contour. (b) The optimal growth temperature contour is shown as a surface plot to illustrate the ethylene/hydrogen relationship with temperature.

CONCLUSIONS

This paper discussed an SVM classifier for finding optimal SWNT growth parameters based on an initial data set of 121 growth experiments. The growth system used is unique, allowing for much quicker experimentation and thus the ability to apply an SVM approach. A classifier accuracy of 94.21% was obtained, and the trained model was applied to a large simulated dataset to determine 90% probability of SWNT growth conditions within each growth parameter. The classifier was also used to predict SWNT growth on 20 new growth experiments validating the SWNT growth regions. Additional experimentation is needed to verify the bounds of the predicted growth regions over more expansive conditions. This method demonstrates a way to create generalized models linking input conditions to growth selectivity using only a limited amount of experimental data. Data mining approaches, and support vector machine classifiers in particular are able to significantly reduce the number of actual experiments needed to explore complex experimental parameter spaces. Having established high-probability regions of selective SWNT growth, we can now explore more deeply, investigating growth rates and chiral dependencies.

REFERENCES

1. M.S. Dresselhaus, G. Dresselhaus and P.C. Ecklund, *Science of Fullerenes and Carbon Nanotubes,* 802 (1996)
2. H. Kataura, A. Kimura, Y. Ohtsuka, S. Suzuki, Y. Maniwa, T. Hanyu and Y. Achiba, Jpn. J, *Appl. Phys.* **37,** L616 (1998)
3. H. Zhu, J. Wei K. Wang D. Wu, K. Suenaga, *J. Crystal Growth,* **310,** 5473 (2008)
4. S. Reich, J. Robertson, L. Li. "Control the chirality of carbon nanotubes by epitaxial growth". *Chemical Physics Letters,* **421,**469–472 (2006)
5. R. Rao, T. Cherukuri B. Yakobson B. Maruyama, D. Liptak. "In situ evidence for chirality-dependent growth rates of individual carbon nanotubes*" Nature Materials,* **11,** 213–216 (2012)
6. M. Grujicic, B. Gersten, G. Cao. "Optimization of the chemical vapor deposition process for carbon nanotubes fabrication*" Applied Surface Science,* **199,** 90–106 (2002)
7. A.A. Puretky, S. Jesse I.N. Ivanov G. Eres, D.B. Geohegan. "In situ measurements and modeling of carbon nanotube array growth kinetics during chemical vapor deposition*" Applied Physics,* **81,** 223–240 (2005)
8. Cortes, Corinna and Vladimir Vapnik. "Support-vector networks". *Machine Learning,* **20(3)** 273–297 (1995)
9. M. Hall, E. Frank, G. Holmes, B. Pfahringer, P. Reutemann, I. Witten. "The WEKA Mining Software: An Update" *SIGKDD Explorations,* **11(1),** (2009)
10. J. Platt. "Fast Training of Support Vector Machines using Sequential Minimal Optimization" *Advances in Kernel Methods – Support Vector Learning.* (1998) (in press)
11. M. Balkanski, R.F.Wallis, and E. Haro "Anharmonic effects in light scattering due to optical phonons in silicon" *Phys Rev B* **28**, 1928 (1983)

Mater. Res. Soc. Symp. Proc. Vol. 1700 © 2014 Materials Research Society
DOI: 10.1557/opl.2014.664

On the Amorphisation Trajectory of Carbon Nanotubes

Saveria Santangelo[1] and Candida Milone[2]

[1]Dipartimento di Ingegneria Civile, dell'Energia, dell'Ambiente e dei Materiali (DICEAM)
Università "Mediterranea", Loc. Feo di Vito, 89122 Reggio Calabria, Italy.

[2]Dipartimento di Ingegneria Elettronica, Chimica ed Ingegneria Industriale (DIECII)
Università di Messina, Viale Ferdinando Stagno d'Alcontres 31, 98166 Messina, Italy.

ABSTRACT

A very simple model for the kinetics of oxidation of carbon Nanotubes (NTs) is proposed which is able to reproduce the main features of their measured kinetic thermal oxidation profiles. Based on this model the resistance to oxidation of NTs is found to decrease with increasing defect density and amorphous phases, i.e. sp^3 bonding component. This finding supports the validity of assumptions previously made to explain the correlation between results of Raman Spectroscopy (RS) and Kinetic Thermal Analysis (KTA) on NTs via a three-stage model, inspired to that proposed by Ferrari and Robertson for other nanocarbons.

INTRODUCTION

Despite of the increasing popularity of other forms of carbon, NTs continue to attract much attention for their great commercial potential in a large variety of nanotechnology applications [1]. RS is widely utilized for the fast and non-destructive characterization of all forms of carbon [2,3]. It allows easily identifying its allotropes and evaluating effects of disorder/ defects that are of great interest in all fields where localization matters.

Very recently, by systematically investigating a wide set of NTs with different specifics (size, morphology, crystalline quality and purity degree), we demonstrated [4] the existence of a correlation between strength of C bonding, as measured by the G-band position (ω_G) in Raman spectra, and oxidative resistance of the samples, as monitored by the maximum oxidation-rate temperature (T_M) in kinetic thermal profiles.

In order to get a deeper insight into the problem, here we elementarily model the NT oxidation process.

CONSIDERED SAMPLES

All the samples considered, listed in Table I, were object of previous studies focused on various aspects of their behavior [4–6]. Samples include Heat Treated (HT) at moderate temperature (1273–1773 K), as well as as-grown NTs. The latter sample typology comprises

Table I. Main properties of the considered samples: metallic impurity content (x_M), amorphous fraction (ξ_{AC}) and oxidative resistance (T_M) of the samples as assessed by TG and KTA, and defect density and strength of C bonding, as monitored by D/G intensity ratio (I_D/I_G) and position of the G-band (ω_G) in Raman spectra.

Sample Code	Ref	x_M (wt%)	ξ_{AC} (wt%)	T_M (K)	I_D/I_G	ω_G (cm^{-1})	Sample Code	Ref	x_M (wt%)	ξ_{AC} (wt%)	T_M (K)	I_D/I_G	ω_G (cm^{-1})
AR1	4	0.3	0.0	811.7	0.41	1582.6	HT1	4	2.7	0.0	856.9	1.78	1579.5
AR2	5	3.3	0.0	815.3	1.41	1584.1	HT2	4	1.7	0.0	865.2	1.34	1579.4
AR3	5	6.7	0.0	843.7	1.18	1575.8	HT3	4	2.7	0.0	912.1	1.78	1582.6
AR4	5	4.1	0.0	823.8	2.20	1576.5	HT4	4	1.7	0.0	892.0	0.99	1583.4
AR5	5	3.7	0.0	839.0	1.40	1575.6	AC1	6	8.0	3.3	820.0	2.12	1581.6
HQG	5	0.0	0.0	1054.9	0.11	1582.0	AC2	6	18.3	1.5	802.6	1.50	1584.4
LQG	5	0.0	0.0	1038.1	0.47	1580.8	AC3	6	12.2	3.3	806.1	2.11	1580.2
AAC	4	2.7	100.0	756.5	0.48	1597.0	AC4	6	4.5	0.4	829.1	1.90	1579.6
AF1	5	17.8	0.0	812.1	1.30	1582.0	AC5	6	1.3	5.2	820.5	1.21	1576.3
AF2	5	3.5	0.0	827.2	1.78	1576.6	AC6	6	4.7	2.9	812.9	1.02	1579.3
AF3	5	7.3	0.0	831.0	1.49	1576.5	AC7	6	4.6	4.4	813.7	2.20	1580.0
AF4	5	15.2	0.0	825.6	1.61	1577.3	AC8	6	2.0	5.0	848.9	0.79	1575.6
AF5	5	11.2	0.0	831.1	1.43	1577.2	AC9	6	1.6	4.9	819.3	2.15	1582.7
AF6	5	4.2	0.0	843.5	0.86	1577.4	AC10	6	1.0	2.8	806.8	1.19	1583.0
AF7	5	11.1	0.0	827.8	0.52	1575.7	AC11	6	1.3	3.6	810.3	1.96	1582.3
AF8	5	10.4	0.0	818.9	1.11	1580.8	AC12	6	1.9	3.2	822.9	0.89	1575.7

both As-Received (AR, from Bayer and Helix Material Solutions Inc) and laboratory-prepared NTs, coded as "AF" or "AC" depending on they were Amorphous carbon Free or Amorphous carbon Containing, respectively. In addition, Amorphous Activated Carbon (AAC) supplied by Carlo Erba, High-Quality Graphite (HQG) purchased from Aldrich and a Lower-Quality Graphite (LQG) sample, obtained by grinding for 1 h the as-received graphite, are regarded as reference samples. All these nanocarbons were systematically investigated by means of RS, KTA, Thermo-Gravimetry (TG) and complementary techniques. An accurate description of results obtained, as well as technical details concerning instrumentation utilized and data analysis can be found elsewhere [4−6].

MODELING OF THE NT OXIDATION PROCESS

Below, we elementarily model the oxidation process of NTs in order to identify the most influential factors on their oxidative resistance.

It is known that oxidizing high-grade NTs for short duration results in the etching away the tube caps and the thinning of the tubes through a layer-by-layer peeling of the outer walls, starting from more strained regions (alike tube caps) [7,8]. Also in the case of low-grade NTs, oxidation at low temperature, commonly adopted for the selective removal of amorphous phases from tube walls [9], leads to the tube thinning, as we previously showed for AC-NTs [6]. This is because also defects along the length of the tubes contribute to decomposition of NTs by providing initiation sites [7,8,10].

Based on these experimental findings, in order to simulate the NT oxidation process, we made the following extremely simple assumptions:

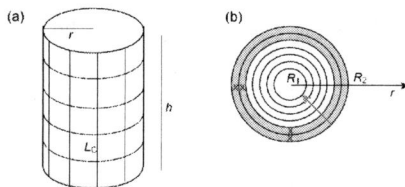

(a)   (b)

Figure 1. (a) Sketch of a tube wall with point defects ideally located at the boundary of the regions of side L_C. (b) Radial section of a NT of inner and outer diameters R_1 and R_2 with external amorphous coating (grey section, where the blue crosses represent sp^3 interconnects between adjacent layers). The red arrow indicates the burning front direction.

    i) If tubes are sufficiently long, the effects due to defects located at the tube ends can be neglected;

    ii) In such a case, the oxidation starts at defects located on the tube walls and progresses in *radial* (rather than in *axial*) direction;

    iii) The local oxidation-rate is proportional to the density of lattice defects on the exposed to oxygen tube walls;

    iv) Metallic impurities play no role because they are passivated by coke [4];

    v) Defects are thought of as located on graphene layers at the boundary of circular regions of diameter L_C, with L_C denoting the average in-plain inter-defect distance, as usually estimated by RS [2] using the D/G intensity ratio ($L_C \propto I_D/I_G^{-1}$). For simplicity, square regions of side L_C can replace the circular ones (without loss of generality). The generic graphene layer of height h and radius r (Fig. 1a) contains a number

$$n_R = \frac{2\pi\, r\, h}{L_C^2}$$

of square regions. If each of them contributes with $2L_C$ point-defects (i.e. proportionally to its boundary length), the total number of defects in the layer is

$$N_D = n_R\, 2L_C = \frac{4\pi\, r\, h}{L_C}.$$

Thus, the defect density per length unit is

$$n_D = \frac{N_D}{h} = \frac{4\pi\, r}{L_C} \propto \frac{r}{L_C},$$

and, based on assumption iii, the burning front progresses in $-\underline{r}$ direction (Fig. 1b) at a local speed $v \propto n_D \propto r/L_C$.

    vi) If an amorphous layer coats the tube, two different L_C values are used, a greater one for the ordered NT core, $L_{C,1}$, and a by far smaller one, $L_{C,2}$, for the external amorphous coating (grey colored in sectional view of Fig. 1b). As the speed at which the oxidation propagates toward the tube axis is inversely proportional to the in-plain inter-defect distance, within the external amorphous coating the oxidation front progresses at higher speed ($v_2 > v_1$), that is the distance, $\Delta r \equiv d$, between two adjacent layers is covered in a shorter time, $t_2 = \Delta r/v_2 < t_1 = \Delta r/v_1$.

    vii) Finally, interconnects between adjacent layers (red crosses in sectional view of Fig. 1b) are envisaged to produce the simultaneous burning of two or more layers ($l \geq 2$), that is the oxidation front propagates in radial direction at higher speed ($v' = lv$ against v), covering a distance, $d' = ld$, multiple of d in the same time as d in absence of interconnects.

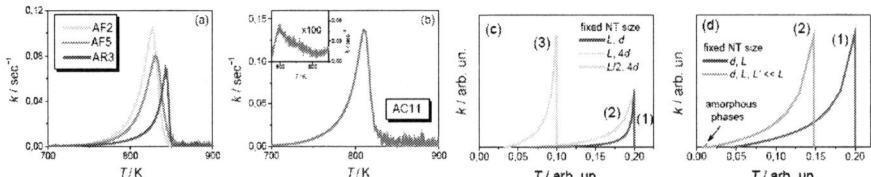

Figure 2. (a–b) Experimental and (c–d) simulated kinetic thermal profiles. Experimental profiles refer to (a) amorphous carbon free NTs with different crystallization degree and (b) amorphous carbon coated NTs. Simulated profiles show effects of (c) in-plane inter-defect distance (L vs $L/2$) and interconnects (d vs $4d$), and (d) presence of amorphous phases.

RESULTS

The rate constant for carbon (and NT) oxidation reaction is typically expressed as $k(T) = -(1/m) \cdot dm/dt$ [11], where m is the instantaneous mass of carbon and T is the temperature linearly increasing with t. Operating under quasi-isothermal conditions [5], the sample temperature can be assumed as constant, so that $k(T)$ can be regarded as the instantaneous reaction rate constant at the given T. Figures 2a and 2b show experimental profiles typically obtained in absence and presence of amorphous phases, respectively. As previously reported [6], the presence of amorphous phases results in a very weak contribution on the lower-T side of the curve, as well as in a lower T_M for given NT diameters.

In order to simulate the oxidation process, the nanotube is envisaged to decompose layer by layer, and the quantity $-(1/m) \cdot dm/dt$ is calculated for uni-modal distributions of inner and outer diameters. Since temperature is increased at constant rate (i.e. $T \propto t$), calculated k values are plotted as a function of $t \propto L_C/r$. Results typically obtained are shown in Figs. 2c and 2d.

In absence of amorphous phases, for fixed NT diameters, the larger L_C, the higher T_M (compare curves 2 and 3 in Fig. 2c), in accord with experimental findings (see samples AF2 and AF5 in Fig. 2a, and ref. [5]); the presence of interconnects reflects on a broadening of the profile (compare curves 1 and 2 in Fig. 2c). When the two effects combine, the profile is broader and peaks at lower temperature (compare curves 1 and 3 in Fig. 2c), as observed experimentally [6].

In presence of amorphous phases, the use of two different L_C values for external amorphous coating and ordered NT core leads to a simulated profile that exhibits a weak contribution on the lower-T side and that, for fixed NT diameters, peaks at lower T_M, as experimentally observed (see samples AR3 in Fig. 2a and AC11 in Fig. 2b, and ref. [6]).

Although much work has to be still done in order to take into account also the effects of tube length and the influence of diameter-distributions, as demonstrated by the comparison between measured and calculated profiles (Fig. 2), the basic modeling here proposed is able to reproduce the *essentials* of the experimentally observed NT oxidation kinetics, proving the properties of the assumptions made.

DISCUSSION

The shown simulation results prove that density and typology of defects mainly influence the oxidative resistance of NTs. In particular, shortening of the in-plane inter-defect distance, and

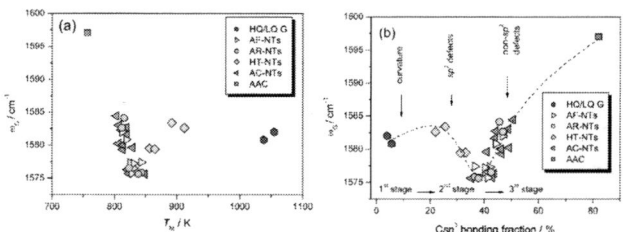

Figure 3. ω_G as a function of (a) T_M and (b) Csp^3 bonding fraction of considered nanocarbons, $-31ln((T_M-780)/440)$, as here empirically estimated from T_M values reported in Table I.

presence of interconnects between adjacent graphene layers and amorphous phases, all contribute to increase the localization of π-electrons, in agreement with literature [7].

Any variation of influential factors leading to an increase of sp^3 C bonding component is found to produce a lowering of T_M. This supports our previous interpretation of the data shown in Fig. 3a. The non-monotonic change of ω_G with T_M reflects the non-monotonic dependence of the strength of C bonding on the variation of the sp^3 C bonding fraction [4], as occurs in other nanocarbons [12].

The sp^2 C bonding component is known to be only 5–10% in carbonized wood powder and carbon-black pellet, 15% in diamond like-carbon and ~50% in SWNTs [13]. The sp^2 C fraction of sample AF2, whose representative point lies around the minimum of the ω_G vs T_M curve (Fig. 3a), is approximately 59% [14]. Annealing of NTs at moderate temperatures produces a limited decrease of sp^3 hybridized C bonds [15]. Hence, assuming oxidative resistance of 40 μm-sized graphite platelets [16] and amorphous carbon [17] as limiting values, T_M can be expressed as a function of the C sp^3-bonding fraction to *coarsely* fit the above cited values. Inverting the found relationship allows *roughly* estimating the C sp^3-bonding fraction if the oxidative resistance is known.

Figure 3b shows ω_G as a function of the C sp^3-bonding fraction of considered nanocarbons, as here empirically estimated from T_M values reported in Table I. It represents the "amorphisation trajectory" of NTs, whose existence we previously proposed to account for the found dependence of ω_G on T_M [4] on the basis of that proposed by Ferrari and Robertson for other nanocarbons [12].

The first stage depicts the effects on C bonding of the introduction of curvature in the graphene, accompanying the evolution from the flat 100% sp^2 structure of graphite to the curved one of well-crystallized HT-NTs with still high sp^2 C bonding component. The second stage describes the further deviation from the ideal graphitic network due to the introduction of sp^2 defects in NTs walls (i.e. distorted hexagons, pentagons and heptagons, responsible for tube twisting and bending) accompanying the evolution from HT-NTs to pristine AR-NTs and AF-NTs that have better crystalline quality (i.e. lower I_D/I_G and $\xi_{AC} = 0$). Finally, the third stage depicts the further increase of the localization of π-electrons due to the introduction of non-sp^2 defects (i.e. edge C atoms and interconnected adjacent layers involving sp^3 hybridized C bonds) in walls of AC-NTs.

13

CONCLUSIONS

The oxidation kinetics of carbon nanotubes is elementarily modeled. Calculated kinetic thermal profiles well reproduce the main features of the measured ones. By this approach, density and typology of the defects (i.e. in-plane inter-defect distance, and presence of amorphous phases and interconnects between adjacent graphene layers) are found to be the factors that chiefly influence the oxidative resistance of NTs. The decrease of resistance to oxidation is due to the increase of the sp^3 bonding component.

This finding supports the property of our previous choice [5] of a three-stage model, inspired to that proposed by Ferrari and Robertson for other nanocarbons [12], to explain the non-monotonic dependence of strength of C bonds, as described by the Raman G-band position, on sp^3 C bonding fraction, as monitored by resistance to oxidation of NTs assessed by KTA.

ACKNOWLEDGMENTS

This work was partly financed by Progetto Operativo Nazionale Ricerca e Competitività 2007–2013 PON 01_01869 (TEMADITUTELA).

REFERENCES

1. *Carbon Nanotubes: Advanced Topics in the Synthesis, Structure, Properties and Applications*, eds. A. Jorio, G. Dresselhaus, and M. S. Dresselhaus, Topics in Applied Physics 111 (Springer-Verlag Berlin Heidelberg, 2008).
2. M. S. Dresselhaus G. Dresselhaus, R. Saito, and A. Jorio, *Phys. Rep.* **409**, 47 (2005).
3. L.M. Malarda, M.A. Pimenta, G. Dresselhaus, and M. S. Dresselhaus, *Phys. Rep.* **473**, 51 (2009).
4. S. Santangelo, and C. Milone, *J. Phys. Chem. C* **117**, 14206 (2013).
5. S. Santangelo, G. Messina, G. Faggio, M. Lanza, and C. Milone, *J. Raman Spectr.* **42**, 593 (2011).
6. S. Santangelo, M. Lanza, and C. Milone, *J. Phys. Chem. C.* **117**, 4815 (2013).
7. J. H. Lehman, M. Terrones, E. Mansfield, K. E. Hurst, and V. Meunier, *Carbon* **49**, 2581 (2011).
8. P. M. Ajayan, T. W. Ebbesen, T. Ichihashi, S. Iijima, K. Tanigaki, and H. Hiura, *Nature* **362**, 522 (1993).
9. S. Osswald, E. Flahaut, and Y. Gogotsi, *Chem. Mater.* **18**, 1525 (2006).
10. D. Bom, R. Andrews, and D. Jacques, *Nano Lett.* **2**, 615 (2002).
11. K. Zaghib, X. Song, and K. Kinoshita, *Termochim. Acta* **371**, 57 (2001).
12. A. C. Ferrari, and J. Robertson, *Phys. Rev. B* **61**, 14095 (2001).
13. A. Mezzi, and S.Kaciulis, *Surf. Interface Anal.* **42**, 1082 (2010).
14. E. Fazio, E. Piperopoulos, S. H. Abdul Rahim, M. Lanza, G. Faggio, G. Mondio, F. Neri, A. M. Mezzasalma, C. Milone, and S. Santangelo, *Curr. Appl. Phys.* **13**, 748 (2013).
15. B. Lesiak, J. Zemek, P. Jiricek, and L. Stobinski, *Phys. Status Solidi B* **246**, 2645 (2009).
16. W. Jiang, G.Nadeau, K. Zaghib, and K. Kinoshita, *Thermochim. Acta* **351**, 85 (2000).
17. W. Huang, Y. Wang, G. Luo, and F. Wei, *Carbon* **41**, 2585 (2003).

Electrical Investigation

Mater. Res. Soc. Symp. Proc. Vol. 1700 © 2014 Materials Research Society
DOI: 10.1557/opl.2014.731

Simulation of Charge Transport in Disordered Assemblies of Metallic Nano-Islands: Application to Boron-Nitride Nanotubes Functionalized with Gold Quantum Dots

John A. Jaszczak[1], Madhusudan A. Savaikar[1], Douglas R. Banyai[1], Boyi Hao[1], Dongyan Zhang[1], Paul L. Bergstrom[2], An-Ping Li[3], Juan-Carlos Idrobo[4], and Yoke Khin Yap[1]

[1]Department of Physics, Michigan Technological University, Houghton, MI 49931, U.S.A.
[2]Department of Electrical and Computer Engineering, Michigan Technological University, Houghton, MI 49931, USA.
[3]Center for Nanophase Materials Sciences, Oak Ridge National Laboratory, Oak Ridge, TN 37831, USA.
[4]Materials Science and Technology Division, Oak Ridge National Laboratory, Oak Ridge, TN 37831, USA.

ABSTRACT

In this study, we investigate the charge-transport behavior in a disordered one-dimensional (1D) chain of metallic islands using the newly developed multi-island transport simulator (MITS) based on semi-classical tunneling theory and kinetic Monte Carlo simulation. The 1D chain is parameterized to model the experimentally-realized devices studied by Lee *et al.* [*Advanced Materials* **25**, 4544-4548 (2013)], which consists of nano-meter-sized gold islands randomly deposited on an insulating boron-nitride nanotube. These devices show semiconductor-like behavior without having semiconductor materials. The effects of disorder, device length, temperature, and source-drain bias voltage (V_{SD}) on the current are examined. Preliminary results of random assemblies of gold nano-islands in two dimensions (2D) are also examined in light of the 1D results.

At $T = 0$ K and low source-drain bias voltages, the disordered 1D-chain device shows charge-transport characteristics with a well-defined Coulomb blockade (CB) and Coulomb staircase (CS) features that are manifestations of the nanometer size of the islands and their separations. In agreement with experimental observations, the CB and the blockade threshold voltage (V_{th}) at which the device begins to conduct increases linearly with increasing chain length. The CS structures are more pronounced in longer chains, but disappear at high V_{SD}. Due to tunneling barrier suppression at high bias, the current-voltage characteristics for $V_{SD} > V_{th}$ follow a non-linear relationship. Smaller islands have a dominant effect on the CB and V_{th} due to capacitive effects. On the other hand, the wider junctions with their large tunneling resistances predominantly determine the overall device current. This study indicates that smaller islands with smaller inter-island spacings are better suited for practical applications. Temperature has minimal effects on high-bias current behavior, but the CB is diminished as V_{th} decreases with increasing temperature.

In 2D systems with sufficient disorder, our studies demonstrate the existence of a dominant conducting path (DCP) along which most of the current is conveyed, making the device effectively quasi-1-dimensional. The existence of a DCP is sensitive to the device structure, but can be robust with respect to changes in V_{SD}.

INTRODUCTION

Recent advances in the development of new materials and fabrication techniques have spurred continued interest in further miniaturization of conventional field-effect devices with new device structure designs [1]. Multi-gate architectures have been fabricated that may allow further reduction in the dimensions of classical metal-oxide semiconductor field effect transistors (MOSFET) without degrading the transistor performance [2]. On the other hand, conduction by tunneling in granular metallic systems has been a subject of interest for many years [3,4]. Single-electron transport devices that operate based on tunneling of individual electrons through junctions formed with one or more nanometer-sized islands have been demonstrated, some even operating at room temperature [5,6]. Successful attempts have been made to demonstrate their use as single-electron memory devices and for nanometer-scale displacement sensing [7,8].

This computational study attempts to complement experimental work seeking to elucidate the effects of different factors such as structural disorder on electron tunneling transport [9,11] by beginning to systematically explore the effects of island sizes, inter-island spacings, and conduction channel length on IV characteristics [10]. Particular focus is given to modeling charge transport in boron-nitride nantotubes (BNNTs) functionalized with nanometer-size gold islands. The device properties are investigated at low and high biases, and the effects of temperature on the Coulomb blockade and the device threshold voltage are studied. Later, the work is further extended to study the effect of structural disorder on 2D device characteristics that gives an insight into the functioning of experimentally fabricated multi-dimensional devices.

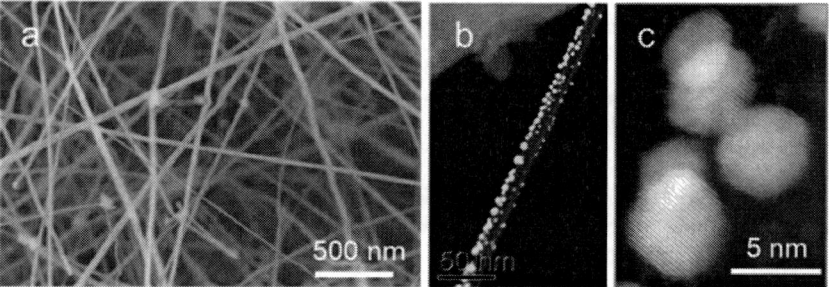

Figure 1. Images of gold quantum dot functionalized boron nitride nanotube s (QDs-BNNTs) obtained by (a) scanning electron microscopy and (b,c) scanning transmission electron microscopy. Reprinted with permission from Lee *et al.* [11]. Copyright © 2013 WILEY-VCH Verlag GmbH & Co. KGaA, Weinheim.

Functionalization of high-quality 20-80 nm diameter BNNTs with gold quantum dots deposited by pulsed laser deposition has recently been demonstrated by Lee *et al.* [11] (figure 1). Without gold-dot functionalization, the BNNTs are excellent insulators, and show currents of less than 10^{-11} A under bias potentials up to 180 V. On the other hand, the gold quantum-dot-functionalized BNNTs (QDs-BNNTs) exhibit room temperature semiconductor-like switching behavior, with turn-on voltages (V_{th}) in the range of 2.0 to 34.0 V, increasing with increasing length (L) of the QDs-BNNT device, where L ranges from 1.29 to 2.37 μm (figure 2).

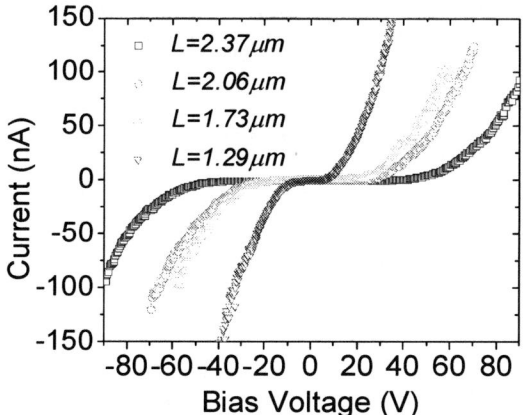

Figure 2. Current-voltage characteristics of QDs-BNNT devices of different lengths demonstrating non-Ohmic behavior, and Coulomb-blockade effects. Data collected using 4-probe scanning tunneling microscopy. Reprinted with permission from Lee *et al.* [11]. Copyright © 2013 WILEY-VCH Verlag GmbH & Co. KGaA, Weinheim.

THEORY AND SIMULATION

Initial investigations were focused on model one-dimensional systems corresponding to the QDs-BNNTs of Ref. 11 (figure 3). The model device consists of a chain of 199 gold islands (200 junctions) between source and drain electrodes. In this study, the BNNT is assumed to play no role other than to geometrically align the islands because of its insulating nature in the absence of gold islands. The radius of each island is randomly selected from a uniform distribution between 3 and 10 nm, while the junction widths are randomly chosen from a uniform distribution in the range of 1 and 5 nm. An island at one end of the chain is selected to be a fixed drain electrode, while the source (ground) electrode is chosen from among the remaining islands in the chain, thus fixing the number of islands in the system (chain) and its length.

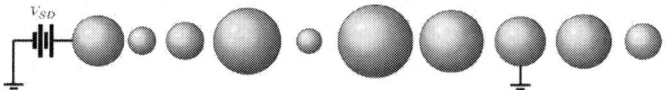

Figure 3. Schematic of the geometrical model of a 1D chain of gold nano-islands with randomly selected island radii and junction widths, deposited on an insulating born-nitride nanotube used for the MITS simulation of systems fabricated by Lee *et al.* [11]. Reprinted with permission from Savaikar *et al.* [10]. Copyright 2013, AIP Publishing LLC.

Conduction in the multi-island devices is modeled using kinetic Monte Carlo simulation methods [12-13] based on tunneling rates that are computed semi-classically (see Ref. 10 for further details). The probabilities for tunneling between any pair of nearby islands at any given time depend on three primary factors: the charge states of the islands, the voltage drops across the junctions, and the junctions' tunneling resistances. All three of these factors can dynamically vary during the simulation. In particular, in contrast to most models that use tunneling resistances that are fixed throughout the simulation, the tunneling resistances in MITS dynamically vary with the voltage drops across the junctions, both due to the applied voltage bias, and the charge states of the capacitively coupled islands.

The semi-classical approach used for calculating the tunneling rates assumes that (i) the energy spectrum of the conductive islands may be considered continuous (ii) the tunneling time is negligible compared to the time between successive tunneling events, and (iii) coherent tunneling events are ignored [14,15]. For a pair of adjacent islands i and j, the tunneling rate is given by [14-16].

$$\Gamma_{ij}(\Delta W_{ij}) = \left(\frac{-\Delta W_{ij}}{e^2 R_{ij}}\right)\left[1 - \exp\left(\frac{\Delta W_{ij}}{k_B T}\right)\right]^{-1}, \tag{1}$$

where ΔW_{ij} is the change in the free energy of the system due to the tunneling event, R_{ij} is the tunneling resistance of the junction, e is the electron charge, k_B is the Boltzmann constant, and T is the temperature. As is clear from Eq. 1, ΔW_{ij} and R_{ij} each play key roles in determining the tunneling rates across the device.

Consider first the change in free energy due to the transition, which is given by $\Delta W_{ij} = -eV_{ij} + E_{c.ij}$. This depends on the potential drop V_{ij} that exists across the junction before the transition. V_{ij} in turn depends on the capacitances of the system, which are fixed, and the charge state of the system, which dynamically evolves. The junction charging energy, $E_{c.ij}$, is the energy required for a single electron to tunnel across the junction between the two uncharged coupled islands, i and j, and depends on all of the capacitances of the system [14-16]. An analytical method employing image charges was used for the calculation of junction capacitances C_{ij} between neighboring islands [18-19], and the dielectric constant of the junction material was taken simply to be 1. Given the self-capacitances and junction capacitances, a capacitance matrix C is constructed that relates Q, a vector composed of the charges on the islands, and V, a vector composed of the island potentials through the matrix equation $Q = CV$ [15]. As the charge state Q of the device changes, the matrix equation is used to solve for the island potentials.

The tunneling rate for a junction is also dependent on the junction's tunneling resistance R_{ij}, which is a strong function of the device geometry as it increases exponentially with the fixed junction separation d_{ij}. R_{ij} also depends strongly on the height of the energy barrier between the two islands that form the junction. The barrier height depends on the work function of the islands ϕ, as well as the potential drop V_{ij} across them. If eV_{ij} remains small compared to ϕ, it has a negligible effect on the barrier height and R_{ij} would be a constant. Under simulation conditions in which all the junction resistances in a given chain remain constant, the device IV characteristics follow a linear behavior for large source-drain voltage biases. However, under high bias conditions, especially where there is a large charge build on some islands, the potential difference between the neighboring islands can be significant compared to ϕ, leading to significant band bending. As a result, the effective barrier height would strongly depend on V_{ij}

and subsequently, R_{ij} would vary significantly with the applied source-drain bias or with the charge state during the course of the simulation. Although a junction's barrier height decreases approximately linearly from one island to the next, in order to simplify the calculations, the tunneling barrier is taken to be of constant height across the width of the junction, but with a reduced height whose variation is given by $\phi_{eff,ij} = \phi - eV_{ij}/2$, a reasonable approximation as long as for each junction $V_{ij} < \phi$ [17]. Thus the tunneling resistances are given by [4]

$$R_{ij} = \left(\frac{h^3}{64\pi^2 m_e e^2}\right)\left(\frac{E_F + \phi_{eff,ij}}{E_F}\right)^2 \frac{\exp(2\alpha k_0 d_{ij})}{\phi_{eff,ij}}\left(\frac{\alpha k_0}{r_a}\right)\frac{1}{G_{ij}}, \qquad (2)$$

where $k_0 = (2\pi/h)(2m_e \phi_{eff,ij})^{1/2}$, h is the Planck constant and m_e is the free electron mass. α is an enhancement parameter that was taken to be 0.115 to set an overall current scale comparable to that measured by Lee *et al.* [11]. Approximate values of E_F and ϕ for gold have been chosen as 5.5 eV and 4.8 eV, respectively. The average radius of the two spherical islands forming the junction is r_a, and d_{ij} is the closest distance between their surfaces (the junction width). G_{ij} is a purely geometrical factor that takes into account the solid angle subtended by one spherical island at the other across the tunnel junction when considering the current flux [4].

Simulations were carried out using a newly-developed set of MATLAB®-based codes called MITS (Multi-Island Transport Simulator) that is described in detail in Ref. 10. Important features of MITS include the following:

- The system is described by a physical model of islands and electrodes, in contrast to using fixed resistances and capacitances in a circuit model.
- The model is applicable from low to reasonably high V_{SD}. Tunneling barrier heights dynamically change with charge state and V_{SD}.
- All islands within a set proximity limit are capacitively coupled to each other.

To begin a simulation, a physical model of a tunneling device is constructed, consisting of spherical metallic islands arranged in one- or two-dimensions, with desired sizes and spacings. For the modeling of the one-dimensional (1D) QDs-BNNT systems, the capacitances are calculated analytically. For two-dimensional (2D) systems, a finite-element-method of calculating the capacitances has been developed in order to account for the important polarization effects of the metallic islands. The circuit-matrix solver builds the capacitance matrix, by which the charging energies for the transfer of a single electron are calculated across all the junctions in a given chain [14-16]. With the given (fixed) electrode potentials and the known island charges (taken to be zero in the initial system configuration), the capacitance matrix is then used to determine the island potentials. The tunneling resistance solver computes the R_{ij} across all the nearest-neighbor junctions. Once all of the relevant parameters in the system are determined, tunneling rates across the junctions are computed. Following the kinetic Monte Carlo method, a particular tunneling event is randomly selected from among the available events, the corresponding transition is carried out, and the time is updated. Using the system's new charge configuration, the potential drops, the tunneling resistances, and the tunneling rates across all the junctions are recalculated, and the process is repeated for large number of time steps until the current through the device reaches a steady state with satisfactory statistical accuracy.

RESULTS

One-dimensional devices

Current-voltage (IV) characteristics at $T = 0$ K for the model 1D device are shown in figure 4. At high biases (figure 4a) the IV characteristics are non-Ohmic and vary as $I \propto (V_{SD} - V_{th})^\zeta$. The exponent ζ is non-universal and varies between 1 and 3, increasing from 1 with decreasing chain length. For a fixed N, ζ also shows a crossover from a lower value at low bias to a higher value at high bias. The Coulomb blockades and Coulomb staircase (CS) structures are shown for different device lengths in figure 4b. The blockade width and associated threshold voltage V_{th} increase with increasing N. The CS structures are also more pronounced for longer devices.

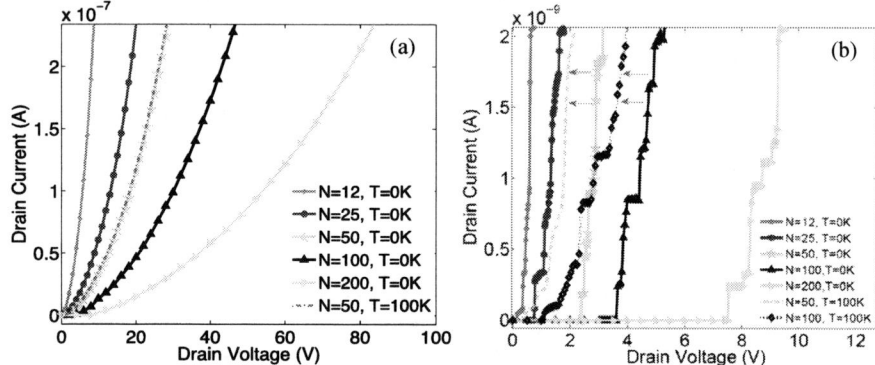

Figure 4. Simulated IV characteristics for the 1D chain of gold islands as a function of N, the number of islands in between the source and drain electrodes at $T = 0$ K. High bias results in (a), which also shows results for $N = 50$ at $T = 100$ K. Low bias results are shown in (b), highlighting the Coulomb Blockade and Coulomb staircase structures at $T = 0$ K and 100 K. Reprinted with permission from Savaikar *et al.* [10]. Copyright 2013, AIP Publishing LLC.

The effects of temperature on the IV characteristics are illustrated in figure 5. As shown in figure 5a, as the temperature is increased, the apparent threshold voltage drops, the Coulomb blockade structure seems to wash out, and the current increases at any given $V_{SD} > V_{th}$. Figure 5b illustrates the effects of temperature on the IV characteristics of a 25-junction device. At low temperatures, the $T = 0$ K turn-on threshold ($V_{th} \approx 0.74$ V) and the Coulomb staircases become rounded. As the temperature is increased, the apparent turn-on threshold voltage, the source-drain bias at which the current reaches some minimum detectable level, decreases. For $T \geq 40$ K, however, additional Coulomb staircase structures manifest themselves at voltages below the $T = 0$ K threshold voltage. For example, at $T = 40$ K a plateau develops in the current for drain voltages between ~0.35 V and 0.55 V. The currents associated with the plateaus of the Coulomb staircase steps also increase in magnitude with increasing temperature, while their widths correspondingly decrease.

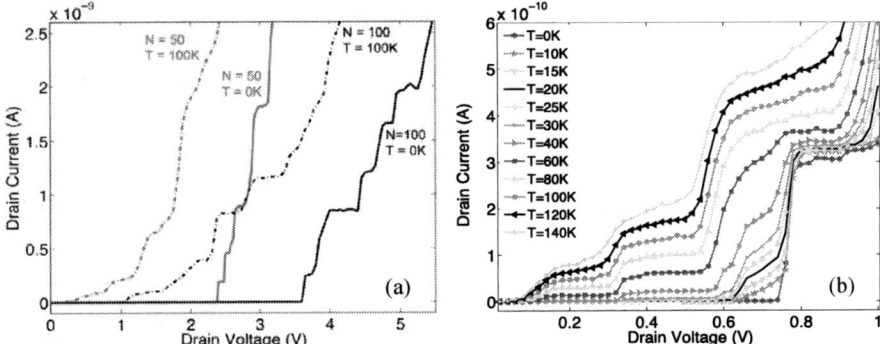

Figure 5. Simulated *IV* characteristics for a 1D chain as a function of temperature. (a) *IV* characteristics for two different chain lengths at $T = 0$ and $T = 100$ K. (b) *IV* characteristics for a device with $N = 25$ junctions at a series of temperatures between 0 and 140 K.

Two-dimensional devices

Experimental studies of 2D systems show *IV* characteristics with similar features to those observed in the 1D systems shown above, including a Coulomb blockade, Coulomb staircases, and non-linear *IV* relationships in the "on" state [4,20-23,26,27]. Preliminary investigations of two-dimensional (2D) random arrays of metallic nano-scale islands were carried out using MITS, and are briefly presented here in order to give a view of capabilities for future work.

Simulations of the 2D systems were carried out on a system of 67 spheres, each of diameter 6.5 nm. Positions of the islands were randomized using Metropolis Monte Carlo, which after decreasing all island diameters to 5.0 nm, resulted in a distribution of nearest-neighbor inter-island spacings ranging between ~1.5 to 5 nm (figure 6), and an average spacing of 2.7±0.1 nm. Simulations were carried out using MITS in the same manner as described above except that the island capacitances and junction capacitances were computed using finite-element methods. The junction capacitances between near-neighbor islands ranged between 1.5×10^{-20} to 10×10^{-20} F. Due to shielding effects from neighboring metallic islands, the island self-capacitances ranged from 0.6×10^{-20} to 4.0×10^{-20} F.

As shown in figure 6, currents tend to flow in the random 2D systems along a fairly narrow dominant conducting path, with many junctions carrying greater than 60% of the total current that is carried to the drain. With increasing V_{SD} (figure 6b) the DCP remained relatively robust, and some junctions in the DCP even increase the fraction of the current they carry. At low but non-zero temperatures (figure 6c), the DCP also remains robust.

The *IV* characteristic for the 67-island 2D device is shown in figure 7 for source-drain biases up to 2 V. The device shows a threshold voltage at ~1 V, and also a weak Coulomb staircase structure compared with the 1D devices. The inset in figure 7 shows the currents as a function of V_{SD} for each individual junction in the DCP. These *IV* curves show weak Coulomb staircase structures reminiscent of the total device *IV* behavior, as one might expect for a junction in a DCP.

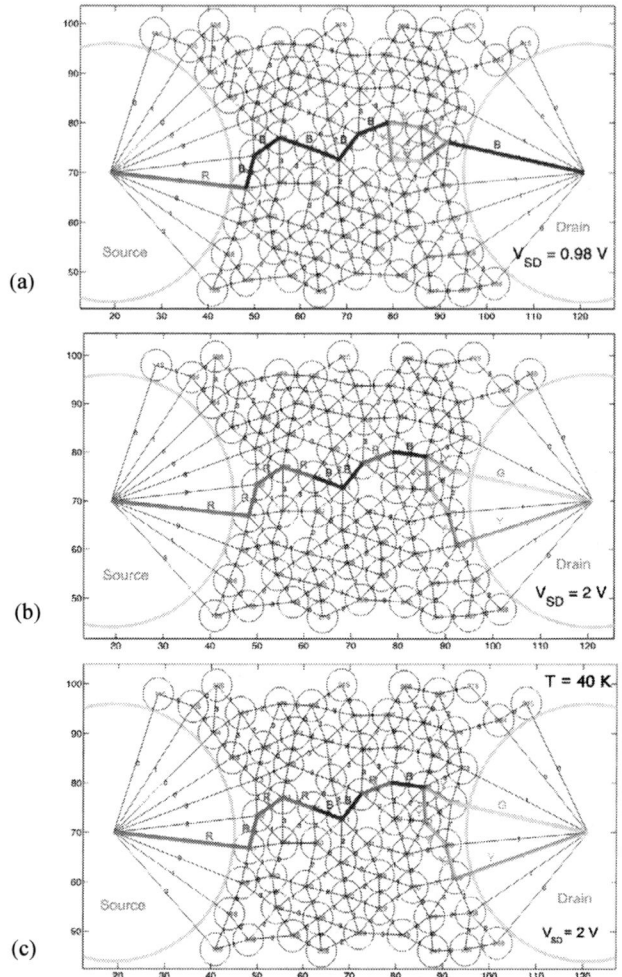

Figure 6. Schematic of a disordered 2D device consisting 67 islands, each of radius of 2.5 nm, randomly positioned on the plane between the source and the drain electrodes (large green ellipses) separated by ~50 nm. The nearest neighbor inter-island spacings range anywhere from ~1.5-5 nm. Allowed current paths are shown as solid black line segments. Junctions carrying significant current are color coded according to the percent of the total current carried to the drain: red (R) = 80-100%, blue (B) = 60-80%, green (G) = 40-60%, yellow (Y) = 20-40%. At T = 0 K as V_{SD} varies from 0.98 V (a) to 2 V (b), the DCP varies but largely retains its dominant conducting nature and position in the 2D array. (c) The same system at V_{SD} = 2V and T = 40K.

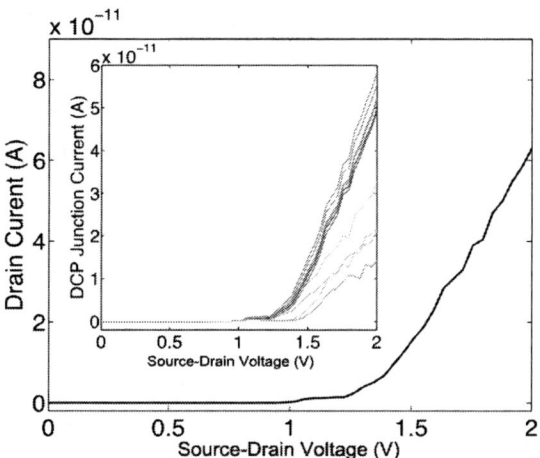

Figure 7. IV characteristics of a 2D device composed of 67 gold islands, as shown in figure 6, at $T = 0$ K, and in the absence of a gate voltage. The inset shows the distribution of currents flowing through individual junctions in the dominant conducting path as a function of applied source-drain voltage bias.

DISCUSSION

At any fixed V_{SD} in the on-state, the 1D devices show decreasing currents with increasing device length, as might be expected due to the increased overall resistance of the longer devices and associated increased number of resistive junctions. However, as demonstrated in figure 8, which shows the variation in the junction resistances as a function of junction width, the wider junctions in a device experience larger voltage drops across them. Because the barrier heights depend on the voltage drops across the junctions, the wider junctions also therefore experience a larger decrease in their tunneling resistances as the source-drain bias is increased (from 12 V to 80V).

MITS simulations demonstrate power-law behavior of the *IV* characteristics for V_{SD} beyond the threshold voltage, consistent with experiments. The non-Ohmic behavior ($\zeta > 1$), in the simulations has been traced to the dependence of the barrier heights on the voltage drops across the junctions, which varies with charge state and with V_{SD}. Whereas Middleton and Wingreen [24] have argued that ζ should equal 1 and 5/3 for infinite 1D and 2D systems, respectively, in the limit of short screening lengths (weak capacitive coupling among islands), their computer simulations for finite systems gave $\zeta = 1$ and 2.0 ± 0.2, respectively. A variety of experimental studies [23,25-28] give exponents ranging between 1 and 3.

Our simulation studies show that the exponent ζ is sensitive to the disorder in the system and the length of the device [10]. The exponents also show a crossover from a lower value to a higher value as the source-drain bias is increased sufficiently. Such crossover behavior has also been observed in experimental devices [25].

25

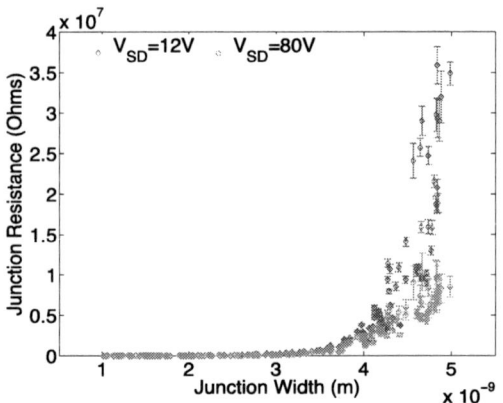

Figure 8. Junction resistances as a function of junction width, for a 200-junction device, for two different source-drain biases, 12 V (blue diamonds) and 80 V (red circles). Note from figure 4 that at 12 V the device is in the Coulomb-staircase regime, while at 80 V it is in the power-law regime. Error bars represent standard deviations in the junction resistances averaged over 5000 Monte Carlo steps (~1 ns at 12 V, and 18 ps at 80 V) after reaching steady state currents. Reprinted with permission from Savaikar *et al.* [10]. Copyright 2013, AIP Publishing LLC.

Values of the threshold voltage V_{th} increase with increasing device length; however, prediction of V_{th} for a device with random island sizes and separations is an open question. Although there is no steady state current for $V_{SD} < V_{th}$, as the applied voltage bias is increased across a device, but below the threshold, the charge state of the system changes in a discrete series of "up-steps" [23,24]. Based on our MITS simulations, these changes in charge state can include the following, alone or in combinations: (i) a change in total charge on the device, (ii) advancement of the charge front across the device, or (iii) rearrangement of charge among the islands. Such transitions occur when V_{SD} is increased sufficiently to bring some ΔW_{ij} to zero making a transition energetically favorable. For example, in a simulation of a 25-junction device, increases in V_{SD} necessary to overcome a total of 27 consecutive up-steps, as V_{SD} is increased from zero to V_{th}, range from 0.01 mV to 153.7 mV. Once a transition takes place, others may follow until once again the system reaches equilibrium. At sufficiently high bias, the last up-step may be overcome, and the ensuing transition will take place with some rate determined by Eq. 1. This transition is a rate-determining step, as a subsequent cascade of transitions then take place quickly, leading to the advancement of one net electron across the device, but ultimately leading to the system coming back to its rate-determining step. Unfortunately, prediction of the individual up-steps and V_{th}, based on a physical model of a 1D random device (materials, island radii, junction separations), appears to be impossible due to the capacitive junction couplings and dependence of the junction voltage drops on the charge state of the system and the applied bias.

With increasing V_{SD} beyond V_{th}, the currents change only slowly due to the slight changes in the junction voltage drops, that is, until some particular V_{ij} reaches a value such that its associated ΔW_{ij} reaches zero and the charge state of the system changes. This can lead to the creation of a new conduction channel (sequence of allowed transitions in charge-state- V_{SD} space that results in a net transfer of charge across the device), leading to a sharp increase in the current and the formation of a step in the CS. With increasing V_{SD}, ever more conduction channels open up, until at sufficiently high V_{SD} the individual CS steps become indistinguishable.

With increasing temperature, key ΔW_{ij} activation barriers for changes in charge state can be thermally overcome that lead to changes in charge state and to non-zero transition rates, even for $V_{SD} < V_{th}$. Thus, thermal effects can lead to a non-linear decrease of the apparent threshold voltage of a device with increasing temperature, and rounding or elimination of steps in the CS structure. New CS steps can even manifest themselves $V_{SD} < V_{th}$ due to the system attaining charge states at non-zero temperatures that are inaccessible at $T = 0$ and $V_{SD} < V_{th}$. Further details of thermal effects in random 1D devices will be the subject of a future publication.

In experimental work on a 2D system of Au grains, by Cordan *et al.* [4] postulated the existence of a quasi-1D dominant conducting path (DCP) that carried most of the current across devices with a wide range of tunneling resistances. Our preliminary work using MITS has demonstrated the existence of a DCP in a random 2D device. The DCP shows robustness with changes in source-drain voltage and with moderate increases in temperature. The Coulomb staircase structure in the 2D device was less pronounced than in comparable 1D systems, however. This is likely due to the DCP being relatively optimized, and thus carrying a narrower range of tunneling resistances along the DCP than which exists among neighboring junctions in the overall 2D device. Our preliminary 2D simulations were also carried out on a relatively small system, leading to a short path length for the DCP. As shown above, the Coulomb staircase structure becomes less prominent as the device length decreases due to shorter devices having a lower probability of sampling unusually large junction widths.

CONCLUSIONS

The MITS simulation package has proven to be a useful tool for modeling *IV* characteristics of 1D and 2D arrays of nano-scale metallic islands under low and high biases, and gaining understanding of underlying mechanisms to explain the Coulomb blockade, Coulomb staircase, and power-law scaling behavior of the devices. The turn-on threshold source-drain bias depends strongly on the capacitances of the system, but because of the inherent randomness in island spacing and radii, the prediction of the threshold voltage based on the physical layout of a random device is not possible without carrying out the full simulation. With increasing source-drain bias the threshold is reached through a series of up-steps in which the charge state of the system changes, and the charge front eventually advances across the device. In agreement with the hypothesis of Cordan *et al.* [4], 2D systems with sufficient disorder have a robust dominant conducting path that carries most of the current across the device.

Future studies are planned to elucidate the effects of temperature and degree of randomness on the behavior of such systems. Such insights may be helpful in using MITS to explore device designs with the goal of engineering desired device characteristics.

ACKNOWLEDGMENTS

Simulation studies were performed in part using the computing cluster wigner.research.mtu.edu in Information Technology Services and rama.phy.mtu.edu in the Department of Physics at Michigan Technological University. Y. K. Yap acknowledges the support from the U.S. Department of Energy, the Office of Basic Energy Sciences (Grant DE-

FG02-06ER46294, PI:Y.K.Y.), the Center for Nanophase Materials Sciences at Oak Ridge National Laboratory (CNMS at ORNL) (Projects CNMS2009-213 and CNMS2012-083, PI:Y.K.Y.), and by ORNL's Shared Research Equipment (ShaRE) User Program (JCI), which are sponsored by the Scientific User Facilities Division, Office of Basic Energy Sciences, U.S. Department of Energy (DOE).

REFERENCES

1. A. M. Ionescu, H. Riel, *Nature* **479**, 329 (2011).
2. I. Ferain, C. A. Colinge, J. P. Colinge, *Nature* **479**, 310 (2011).
3. P. Sheng & B. Abeles, Phys. Rev. Lett., 28, 34 (1972)
4. A. S. Cordan, A. Goltzene, Y. Herve, M. Mejias, C. Vieu, and H. Launois, *J. Appl. Phys.* **84**, 3756 (1998).
5. V. Ray, R. Subramanian, P. Bhadrachalam, L. C. Ma, C. Kim, and S. J. Koh. *Nat. Nanotechnol.* **3**, 603 (2008).
6. P. S. Karre, P. L. Bergstrom, G. Mailick, and S. P. Karna, *J. Appl. Phys.* **102**, 024316 (2007).
7. K. Yano, T. Ishii, T. Hashimoto, T. Kobayashi, F. Murai, and K. Seki, *IEEE Trans. Electron Devices* **41**, 1628 (1994).
8. R. G. Knobel and A. N. Cleland, *Nature* **424**, 291 (2003).
9. R. Parthasarathy, X.-M. Lin, and H. M. Jaeger, *Phys. Rev. Lett.* **87**,186807 (2001).
10. M. A. Savaikar, D. Banyai, P. L. Bergstrom, and J. A. Jaszczak. (2013) Simulation of charge transport in multi-island tunneling devices: Application to disordered one-dimensional systems at low and high bias. *J. Appl. Phys.* **114**, 114504-1-12.
11. C. H. Lee, M. A. Savaikar, J. S. Wang, B. Y. Hao, D. Y. Zhang, D. Banyai, J. A. Jaszczak, and Y. K. Yap. (2013) Room Temperature Tunneling Behaviors of Boron Nitride Nanotubes Functionalized with Gold Quantum Dots. *Advanced Materials* **25**, 4544-4548.
12. A. B. Bortz, M. H. Kalos, and J. L. Lebowitz, *J. Comput. Phys.* **17**, 10 (1975).
13. M. Kotrla, Comp. *Phys. Comm.* **97**, 82 (1996).
14. K. K. Likharev, *Proc. IEEE.* **87**, 606 (1999).
15. C. Wasshuber, *Computational Single-Electronics* (Springer Wien New York, 2001).
16. D. V. Averin and K. K. Likharev. In: *Mesoscopic Phenomena in Solids*, ed. B. Altshuler *et al.* (*Elsevier, Amsterdam*, 1991) p. 173.
17. J. G. Simmons, *J. Appl. Phys.* **34**, 1793 (1963).
18. E. Pisler, and T. Adhikari, *Physica Scipta.* **2**, 81 (1970).
19. J. Lekner, *J. Electrostatics* **69**, 11 (2011).
20. D. D. Cheam, Ph.D. dissertation. Michigan Technological University, Houghton, MI, 2009.
21. A. J. Quinn, M. Biancardo, L. Floyd, M. Belloni, P. R. Ashton, J. A. Preece, C. A. Bignozzi, and G. Redmond. *J. Mater. Chem.* **15**, 4402 (2005).
22. C. A. Neugebauer and M. B. Webb, *J. App. Phys.* **33**, 74 (1962).
23. R. Parthasarathy, X.-M. Lin, K. Elteto, T. F. Rosenbaum, and H. M. Jaeger, *Phys. Rev. Lett.* **92**, 076801 (2004).
24. A. A. Middleton and N. S. Wingreen, *Phys. Rev. Lett.* **71**, 3198 (1993).
25. A. N. Aleshin, H. J. Lee, S. H. Jhang, H. S. Kim, K. Akagi, and Y. W. Park, *Phys. Rev. B* **72**, 153202 (2005).
26. A. J. Rimberg, T. R. Ho, and J. Clarke, *Phys. Rev. Lett.* **74**, 4714 (1995).
27. A. Bezryadin, R. M. Westervelt, and M. Tinkham, *Appl. Phys. Lett.* **74**, 2699 (1999).
28. V. V. Deshpande, M. Brockrath, L. I. Glazman, and A. Yakoby, *Nature* **464**, 209 (2010).

Mater. Res. Soc. Symp. Proc. Vol. 1700 © 2014 Materials Research Society
DOI: 10.1557/opl.2014.675

Interaction volume of electron beam in carbon nanomaterials:
A molecular dynamics study

Masaaki Yasuda, Shinya Wakuda, Yoshiki Asayama, Hiroaki Kawata and Yoshihiko Hirai
Osaka Prefecture University, Sakai, Osaka 599-8531, Japan

ABSTRACT

A molecular dynamics (MD) simulation was performed to study the interaction volume of electron beam in carbon nanomaterials. The interaction between incident electron and carbon atom in the target materials during electron irradiation is introduced by the relativistic binary collision theory. The motion of each atom in the material under electron irradiation is calculated with the MD simulation. The primary energy dependence of the interaction volume in the carbon nanotube and the multi-layered graphene are studied. The secondary damages caused by the knock-on atoms are also discussed.

INTRODUCTION

Irradiating carbon nanomaterials with energetic electrons is expected to become a technique to tailor the structure with desirable properties [1-3]. Cutting [4], bending [5] and welding [6] of carbon nanotubes by electron beam irradiation has been demonstrated. Graphene nanostructures sculpted by electron beam irradiation have been reported [7]. Pill shaped capsules were produced by coalescing C_{60} molecules under electron irradiation [8]. However, such structural modifications of carbon nanomaterials with electron beams are not well established at present. Understanding the atomic level behavior of the materials under electron irradiation becomes important.

Molecular dynamics (MD) simulation is suitable to understand the atomic level behavior of the materials. Several MD simulations have been used to study irradiation-induced defects in carbon nanomaterials [9, 10] and have demonstrated the structural modification of nanomaterials with electron beam [11]. In those studies, a continuous random atom extraction [9] or atom ejection with constant velocity [10, 11] has been introduced as the electron irradiation effects. Electron knock-on cross section [12] and the displacement energy [13] of carbon atoms in single-walled carbon nanotubes (SWCNTs) have been also precisely calculated from the MD simulation. However, an electron bombardment process has not been explicitly included into a simulation of the deformation of carbon nanomaterials.

In our previous studies, we presented MD simulations including electron irradiation effects based on a Monte Carlo method to study the deformation process of carbon nanostructures under electron irradiation. We investigated the structural changes of SWCNTs [14, 15] and graphene [16] caused by electron irradiation. We have reported the changes of the mechanical properties of electron-irradiated SWCNTs [17]. We also studied the correlation between electron irradiation defects and applied stress in SWCNTs [18].

In order to achieve precise modifications of the nanostructures, the control of the interaction volume of electron irradiation becomes important. Because the mean free path of the high energy electron in the carbon nanostructure is enough long, the multiple scattering of the electrons in the structure is neglected. However, even though the electron beam is finely focused, the structural changes are caused in the wider region than the beam spot because of the secondary damages caused by the knock-on atoms. In the present work, we study the interaction volume of electron beam in carbon nanomaterials including the secondary damages caused by the knock-on atoms with the MD simulation.

SIMULASION

Figure 1 shows the simulation model. The electron irradiation effects are introduced using a previously reported model [16, 18]. The collision carbon atom is randomly selected in the carbon nanomaterials. Interaction between an incident electron and the carbon atom is introduced by the Monte Carlo method. The scattering angle of the incident electron ω is determined by the Mott cross-section [19]. Then, the transferred energy from the electron to the target carbon atom E_t and the scattering angle of the carbon atom θ are obtained by the relativistic binary collision model:

$$E_t = \left\{ 1 - \left(\frac{m_e \cos \omega + \sqrt{m_c^2 - m_e^2 \sin^2 \omega}}{m_e + m_c} \right)^2 \right\} \frac{\left(E + 2m_e c^2 \right)}{2m_e c^2} E, \qquad (1)$$

$$\cos \theta = \sqrt{\frac{\left(m_e + m_c \right)^2}{4m_e m_c} \cdot \frac{E_t}{E}}, \qquad (2)$$

where m_e is the electron mass, m_c is the mass of the carbon atom, c is the velocity of light, and E is the incident electron energy. The scattering angle of the carbon atom around the incident axis ϕ is uniformly distributed.

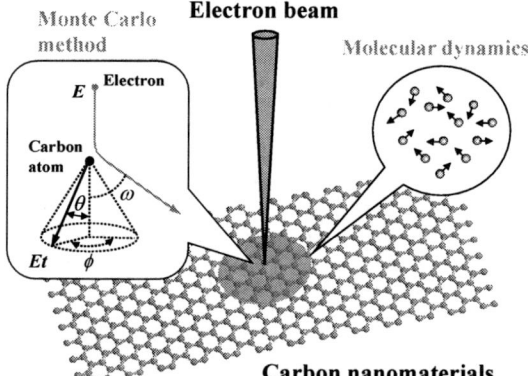

Figure 1. Present simulation model. The structural changes of carbon nanomaterials under electron irradiation were calculated using MD simulations. The collisions of the electrons are introduced by the Monte Carlo method.

The motion of each atom in the carbon nanomaterials under electron irradiation is calculated using MD simulations with empirical potentials. The short- and long-range interactions were described by the Tersoff-Brenner [20-22] and Lennard-Jones potentials, respectively. Both potentials are smoothly connected with cubic splines [23]. The temperature of the nanomaterials under electron irradiation was kept constant at 300 K. Electron irradiation was performed every 2000 MD time steps. The time step used in the MD simulations was 0.5 fs.

As we estimated in our previous study [14], the current density of electron beam in our simulation is about 10 orders of magnitude larger than that in the transmission electron microscope (TEM) experiment. However, the ratio of the ejected carbon atoms per incident electron in the present simulation is independent of the collision rates. Because the cross section of the collision that causes a serious structural change is quite small, the target material is sufficiently relaxed between the irradiation steps. Comparison between the simulations and TEM experiment results is possible by estimating the electron collision number from beam current, the irradiation time and the collision cross-section.

DISCUSSION

At the first setout, we check the energy dependence of the electron irradiation damages of the carbon nanomaterials with the simulation. Figure 2 shows the structures of the (5,5) SWCNTs after electron irradiation at several electron energies obtained with the simulation. The tube length is 5 nm. Both ends of the nanotubes are fixed. The center 2-nm-long region is irradiated by the electrons. The collision rate and the irradiation time are 100 electrons/ps and 5 ns, respectively. The irradiation damages are observed above the electron energy of 95 keV in the present simulation. With an increase in the electron energy, the irradiation damages of the SWCNTs becomes serious.

Figure 2. The structures of (5,5) SWCNTs after electron irradiation at several electron energies.

Figure 3 shows the structures of the (5,5)/(10,10)/(15,15) triple-walled carbon nanotubes (TWCNTs) after focused electron beam irradiation at 300 and 500 keV. The tube length is 5 nm. The center part of the TWCNT is irradiated by the electrons. The diameter of the beam spot is 1 nm. The shaded regions in the figure are irradiated by the electrons. The collision rate and the irradiation time are 20 electrons/ps and 350 ps, respectively. The crosslinks between the tube walls are observed as a typical irradiation damage structure. The irradiation damage at 500 keV is larger than that at 300 keV. The damage is observed outside the irradiated region at 500 keV irradiation.

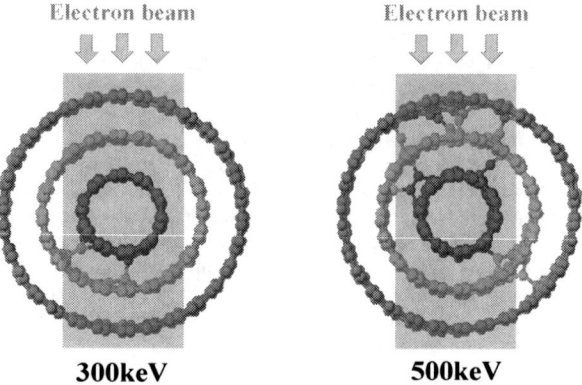

300keV   **500keV**

Figure 3. The structures of (5,5)/(10,10)/(15,15) TWCNTs after focused electron beam irradiation at 300 and 500 keV. The diameter of the beam spot is 1 nm.

        Figure 4 shows top views of the five-layered graphene irradiated by the focused electron beam at several energies. All the layers are irradiated by the electrons. The beam diameter is 1 nm and the size of the beam spot is also shown in the figure. The collision rate and the irradiation time are 50 electrons/ps and 1 ns, respectively. The damaged areas are shown by the circles. The diameters of the damaged areas are 1.13, 1.65 and 1.98 nm for 150, 300 and 500 keV electron irradiation, respectively. With an increase in the electron energy, the damaged area becomes large. Because the mean free path of the high energy electron in the nanostructure is enough long, the multiple scattering of the electrons is not introduced in the simulation. Therefore, the lager damaged area than the beam spot is attributed to the secondary damages caused by the knock-on atoms.

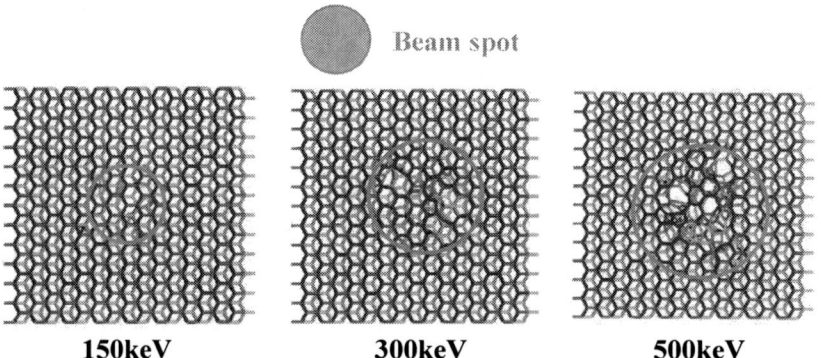

150keV   **300keV**   **500keV**

Figure 4. Top views of the five-layered graphene irradiated by the focused electron beam. The size of the beam spot is also shown.

Figure 5 shows the side view of the five-layered graphene irradiated by electrons at 300 keV. Only the top layer of the graphene is irradiated by the electrons to evaluate the secondary damages caused by the knock-on atoms. The collision rate and the irradiation time are 100 electrons/ps and 50 ps, respectively. The crosslinks between the layers are formed by the knock-on atoms from the upper layers. Although the electrons collide with only the top layer, crosslinks reach fourth layer.

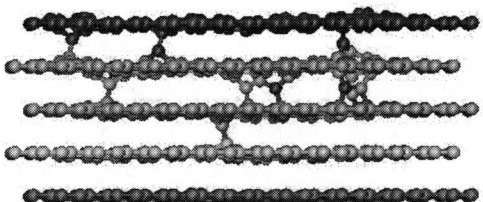

Figure 5. Side view of the five-layered graphene irradiated by electrons at 300 keV. Only the top layer of the graphene is irradiated by the electrons to evaluate the secondary damages caused by the knock-on atoms.

Figure 6 shows the number of damaged layers in electron-irradiated multi-layered graphene as a function of electron energy. Only the top layer of the graphene is irradiated by the electrons. The collision rate and the irradiation time are 100 electrons/ps and 50 ps, respectively. The simulation is performed for 40 times to reduce the statistical fluctuation. The crosslink between the layers are observed above 100 keV. With an increase in the electron energy, the secondary damages reach deeper layer.

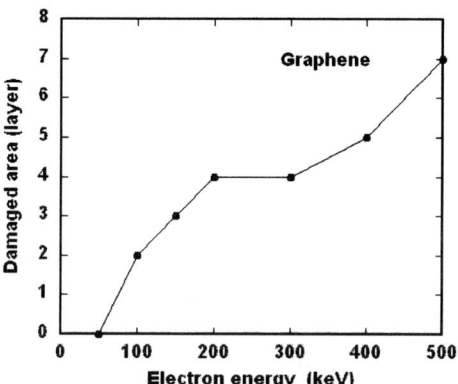

Figure 6. Number of damaged layers in electron-irradiated multi-layered graphene as a function of electron energy. Only the top layer of the multi-layered graphene is irradiated by the electrons.

CONCLUSIONS

    Interaction volume of electron beam in carbon nanomaterials was studied with an MD simulation. The energy dependence of the electron irradiation damages in the carbon nanotubes and graphenes were calculated. With an increase in the electron energy, the damaged area becomes large. This broadening of the damaged area is attributed to the secondary damages caused by the knock-on atoms.

ACKNOWLEDGMENTS

    This work was supported by JSPS KAKENHI Grant number 25249052.

REFERENCES

1. A. V. Krasheninnikov and F. Banhart, *Nat. Mater.* **6**, 723 (2007).
2. A. V. Krasheninnikov and K. Nordlund, *J. Appl. Phys.* **107**, 071301 (2010).
3. G. Gerasimov, *Radiation Synthesis of Materials and Compounds* (CRC Press, 2013), Chapters 18 and 21.
4. F. Banhart, J. Li, and M. Terrones, *Small* **1**, 953 (2005).
5. J. Li and F. Banhart, *Nano Lett.* **4**, 1143 (2004).
6. M. Terrones, F. Banhart, N. Grobert, J.-C. Charlier, H. Terrones, and P. M. Ajayan, *Phys. Rev. Lett.* **89**, 075505 (2002).
7. B. Song, G. F. Schneider, Q. Xu, G. Pandraud, C. Dekker and H. Zandbergen, *Nano Lett.* **11**, 2247 (2011).
8. D. E. Luzzi and B. W. Smith, Carbon **38**, 1751 (2000).
9. P. M. Ajayan, V. Ravikumar, and J.-C. Charlier, *Phys. Rev. Lett.* **81**, 1437 (1998).
10. S. K. Pregler and S. B. Sinnott, *Phys. Rev. B* **73**, 224106 (2006).
11. I. Jang, S. B. Sinnott, D. Danailov, and P. Keblinski, *Nano Lett.* **4**, 109 (2004).
12. A. Zobelli, A. Gloter, C. P. Ewels, G. Seifert, and C. Colliex, *Phys. Rev. B* **75**, 245402 (2007).
13. A. V. Krasheninnikov, F. Banhart, J. X. Li, A. S. Foster, and R. M. Nieminen, *Phys. Rev. B* **72**, 125428 (2005).
14. M. Yasuda, Y. Kimoto, K. Tada, H. Mori, S. Akita, Y. Nakayama, and Y. Hirai, *Phys. Rev. B* **75**, 205406 (2007).
15. M. Yasuda, R. Mimura, H. Kawata, and Y. Hirai, *J. Appl. Phys.* **109**, 054304 (2011).
16. Y. Asayama, M. Yasuda, K. Tada, H. Kawata, and Y. Hirai, *J. Vac. Sci. Technol. B* **30**, 06FJ02 (2012).
17. K. Tada, M. Yasuda, T. Mitsueda, R. Honda, H. Kawata, and Y. Hirai, Microelectron. Eng. **107**, 50 (2013).
18. M. Yasuda, Y. Chihara, K. Tada, H. Kawata, and Y. Hirai, *J. Vac. Sci. Technol. B* **31**, 06FF06 (2013).
19. N. F. Mott, *Proc. R. Soc. (London)* A **124**, 425 (1929).
20. J. Tersoff, *Phys. Rev. B* **37**, 6991 (1988).
21. J. Tersoff, *Phys. Rev. B* **39**, 5566 (1989).
22. D. W. Brenner, *Phys. Rev. B* **42**, 9458 (1990).

23. D. W. Brenner, D. H. Robertson, M. L. Elert, and C. T. White, *Phys. Rev. Lett.* **70**, 2174 (1993).

Mater. Res. Soc. Symp. Proc. Vol. 1700 © 2014 Materials Research Society
DOI: 10.1557/opl.2014.537

Novel method of electrical resistance measurement in structural composite materials for interfacial and dispersion evaluation with nano- and hetero-structures

Joung-Man Park[1], Dong-Jun Kwon[1], Zuo-Jia Wang[1], Joon-Hyung Byun[2], Hyung-Ik Lee[3], Jong-Kyoo Park[3] and Lawrence K. DeVries[4]
[1]School of Materials Science and Engineering,
Gyeongsang National University, Jinju 660-701, Korea
[2]Korea Institute of Materials Sciences, Composites Research Center, Changwon, Korea
[3]4 R&D Center, Agency of Defense Development, Daejeon, Korea
[4]Department of Mechanical Engineering, The University of Utah,
Salt Lake City, Utah 84112, U.S.A.

ABSTRACT

Interest in development in the use of nanoparticles in structural composites for the improvement of thermal conductivity, mechanical properties and electrical properties has recently stimulated some research efforts. Such improvements require the introduction of functional groups and the proper selection and concentration of the nanoparticles, as well as their uniform dispersion. The identification and verification of uniformity of dispersion is very important in the efficient processing for improved performance. Recently, new methods for studying and evaluating the interfacial properties between the reinforcing fibers and the epoxy matrix, have been developed. Distinct from FE-SEM observation, electrical resistance methods are being developed which can be applied for to measure interfacial shear strength (IFSS) and degree of dispersion. The main principle, on which the electrical resistance measurement is based, is Kirchhoff's laws, which considers conductive materials as electrical circuits. In this research, the self sensing character of the conductive carbon nanotubes (CNT) and conventional carbon reinforcing fibers has been successfully used as a method for evaluating the dispersion of nanoparticles and interfacial adhesion. The electrical resistance in these composites was observed to be dependent on differences in wetting and interfacial adhesion between matrix and fillers. In summary, a correlation was observed between the electrical resistance and dispersion and degree of cure. It is felt that these methods, along with the electro-micromechanical methods, provide valuable tools for investigating the role of interfacial behavior on thermal conductivity, electrical and mechanical properties. Optical observations by FE-SEM of degree of dispersion and interfacial adhesion are consistent with the electrical resistance results. Additionally, it may be possible to use electrical resistance circuit analysis to detect the location of and extent of micro-damage within composite materials.

INTRODUCTION

The use of carbon fiber reinforced plastic (CFRP) is rapidly increasing in aerospace and automobile applications. This is in large part due to the increase strength and modulus of carbon fiber composites compared to other materials such as glass fiber composites [1-3]. The fiber-matrix interface has been found to be important for high strength in composites. Methods that have been studied and developed to improve interfacial property of composites include: addition of coupling agents, surface modifications, chemical reaction control and use of nano fillers for reinforcement [4-7].

A large number of mechanical test methods have been used to evaluate the interfacial properties of composites, including macroscopic methods [interlaminar shear stress (ILSS) measurements] and microscopic methods [interfacial shear stress (IFSS) measurements] [8-10]. These two evaluation methods have the advantage of yielding easily interpreted interfacial bonding force results but on the other hand they do consume a great amount of time in sample manufacture, preparation and testing.

Thermodynamics approaches, such as contact angle measurements, which are very different from the mechanical evaluation methods just discussed, have also been used to study interfacial properties. In these tests the static contact angle between different liquid types (polarity, non-polarity solvents) and a solid surface was measured for use in calculating surface energies, work of adhesion and spreading coefficient. Contact angle measurements have the advantage of providing a non-destructive method of interfacial evaluation of the interface between a matrix and reinforcement. It has been hypothesized that the work of adhesion is proportionate to the mechanical strength of an interface [11-13]. However, the calculation processes between the measured contact angles and other interfacial properties are complex, difficult and frequently not well understood.

Rapid, simple and accurate interfacial evaluation methods would be of considerable utility for industrial applications. For this reason, an electrical resistance method, based on Kirchhoff's law, has recently been developed and used for such studies [14]. During mechanical loading the shape of carbon fibers, in a composite, is altered with a concomitant change in electrical resistance. Carbon fibers are used as sensing elements in CFRP due to their electrical conductivity. In some more advanced studies, in CRFP, the carbon fibers were used for both strain sensing and damage sensing. The basic theory, in such studies, was that the movement of carbon fiber could be detected by electrical resistance measurement. During loading, the extent of contact between carbon fibers in a CF tow is altered, and the electrical resistance changes from its initial value [15-17].

In this work, interfacial properties of carbon fiber/polymer composites were evaluated by such electrical resistance measurements. During the wetting process of CF tows by the polymer resins, the change in electrical resistance was recorded and analyzed. The test results for different carbon fibers and polymer resins exhibited different trends. The electrical resistance change was associated with the different wetting conditions, and the interfacial properties were predicted by these electrical resistance measurements. The accuracy of predictions based on the electrical resistance method during polymer resin wetting on CF tow was then assessed by IFSS and ILSS tests.

EXPERIMENT

Materials

In this study two types CF tows, 12 K T700S (Toray Co Ltd., Japan) and 12K T50S-15L (Mitsubishi Co Ltd., Japan), were used as reinforcement materials. Phenolic resol resin (Monsanto Co Ltd., U.S.A.), epoxy resin Bisphenol A type YD-114, YD-128 and bisphenol F type epoxy YDF-175 (Kukdo chemical Co Ltd., Korea) were used as matrix resins. Hardener for the first two epoxy resins was acid anhydride type KBH-1089, and the hardener for YDF-175 was polyamide type G0331.

Experimental setup

CF tows were wet by two different polymeric resins. Figure 1 shows schematically the wetting process of the tows by the resin. For each specimen, 1g of polymeric resin was dropped on the tow. The resin wetting process was observed by reflecting microscope during a 20 min observation period, and subsequently the different wetting conditions were investigated and compared by contact angle measurements. The length of the CF tows was 5 cm, and the gauge length for electrical resistance measurement [using a 4 wire method (34401A multi-meter, Agilent Tech., U.S.A.)] of the CF tow was 2 cm.

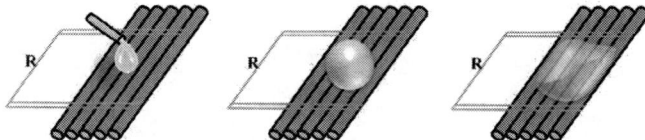

Figure 1. Wetting process of polymer resin on fiber tow

ILSS evaluation was performed based on ASTM D-2344 [18]. Short beam test specimens were manufactured by hand lay-up with a length of 10 mm, a width of 5mm and a thickness of 2.8 mm. Equation (1) was used to calculate the ILSS:

$$ILSS = \frac{3F}{4bd} \qquad (1)$$

where F is the measured maximum load, b is the width, and d is the thickness of the specimen.

The IFSS was determined by the microdroplet pull-out test [19]. The embedded length of resin droplet on a single carbon fiber was 150 μm. The IFSS was calculated from the measured pullout force, F, using the following equation:

$$IFSS = \frac{F}{\pi D_f L} \qquad (2)$$

where D_f and L are the diameter and the embedded length of the fiber in the matrix. A UTM was used to "pull-out" the resin drop on the carbon fiber at a test speed of 1 mm/min.

DISCUSSION

Wetting behavior for electric resistance measurement

The wetting process of polymer resin on CF tow was investigated by a reflective microscope. Figure 2 shows photographs of wetting angle micro-droplets comparing the of wetting processes for phenolic resin and YDF-175 epoxy resin illustrating the very different wetting patterns exhibited by the two resins. The phenolic resin exhibited much better wettability on the CF tow in that the CF tow was nearly completed wetted by phenolic resin in 1 min. The epoxy YDF-175 exhibited much poorer wettability of the CF tow, in that the contact angle of the epoxy resin only decreased from 111° to 92° in 1 min. The wettability of different polymer resin was confirmed by contact angle measurements, in that the phenolic resin exhibited significantly better wettability in comparison to the epoxy resins.

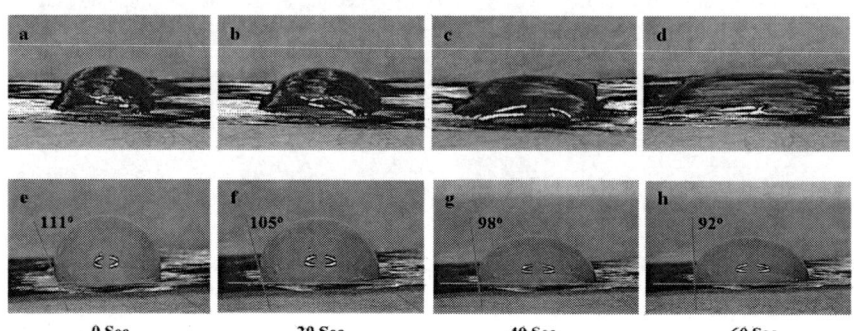

Figure 2. Contact angle measurement of two polymer resins on CF tow (T700S): phenolic resin (a,b,c,d), Bisphenol F type epoxy (e,f,g,h)

Figure 3 shows a schematic model that might be used to explain the differences in wetting behavior for the polymer resins on CF tow surfaces. With good wettability the carbon fibers would tend to be better and more separated by the polymer resin. On the other hand, with poor wettability, much less of the CF tow would be wetted and separated by the resin. It is, therefore, anticipated that the interfacial properties between carbon fibers in a tow and a polymeric resin could be evaluated by such wettabiliy measurements at normal pressure and temperature. The electrical resistance of a CF tow would be influenced by the rearrangement of the carbon fibers during the wetting processes. Conversely, it might be anticipated that wettability might be evaluated by the electrical resistance measurement, in that better wettability between CF tow and polymer resin would be associated with larger electrical resistance changes.

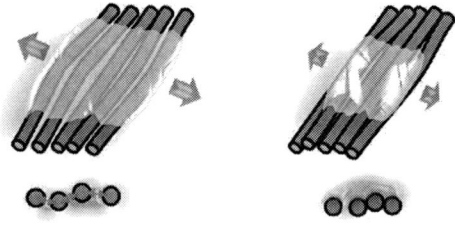

Good wetting   **Poor wetting**
Figure 3. Model of test system for electrical resistance measurement

Figure 4 shows the surface morphology of carbon fibers and the chemical component by IR test. FT-IR results of two kinds of carbon fibers were nearly same. C-C and C=C peaks proved the hydroxyl groups on fiber surface. The chemical analysis was supposed to confirm the difference of sizing on fiber surface, whereas the chemical component of two carbon fibers was similar. However, the surface morphology of carbon fibers was different. The surface of Toray fibers had clear ridges and striations parallel to the fiber axial direction as a result of the carbon fiber manufacturing process. The Mishubithi fiber only exhibited slight ridges. It was predicted that the smoother surface would provide better wetting effect with resins for Mishubithi fiber.

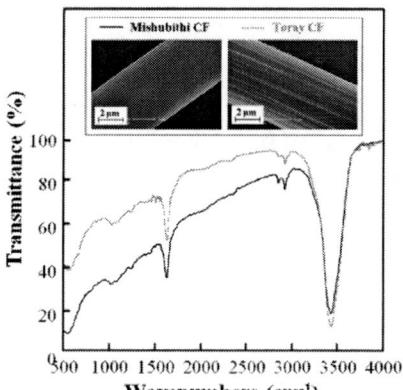

Figure 4. Comparison of chemical and morphology of carbon fibers

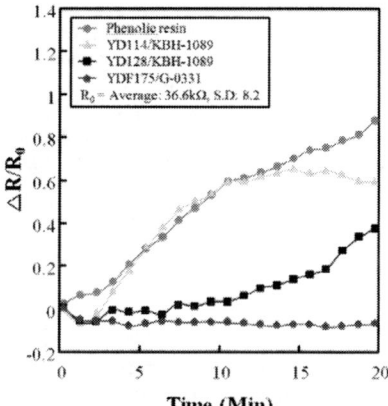

Figure 5. Electrical resistance change during wetting process of polymer resin on carbon fiber (T700S)

Figure 5 shows measured electrical resistance change versus time for different polymeric resins on a T700S CF tow. The phenolic resin exhibited largest electrical resistance change during the 20 minutes observation, indicating that the phenolic resin had the best wettability of the CF tow among the four polymer resins. The electrical resistance change of bisphenol A type epoxy resin cured with acid anhydride hardener was larger than that for the bisphenol F type epoxy resin and polyamide type hardener. The epoxy YDF-175 exhibited smallest electrical resistance change, which is attributed to poorer wetting.

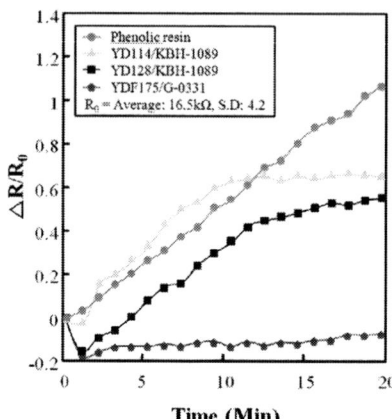

Figure 6. Electrical resistance change during wetting process of polymer resin on carbon fiber (T50S-15L)

Figure 6 shows results similar to those in figure 5 but between T50S-15L CF tow and the polymeric resins. While the electrical resistance results are somewhat similar to those for the T700S CF tow, the electrical resistance changes for some of the resins on the T50S-15L CF tow were higher than on T700S CF tow, which is attributed to the differences in wettability of the carbon fibers by these resins. Phenolic resin which has good wettability on the carbon fibers would tend to be better and more separated by the polymer resin. This is the reason that it exhibited the largest change in electrical resistance during wetting test. The On the other hand, YDF 175 has poor wettability on carbon fibers, and much less of the CF tow would be wetted and separated by the resin. It also exhibited the smallest change in electrical resistance during the wetting test.

Mechanical test of interfacial adhesion

Figure 7 shows the IFSS results of carbon fibers for the different polymeric resins. The interfacial properties of the two types of CF tow and polymeric resins were different, and these results exhibited the same basic trends as the electrical resistance measurements. The T50S-15L carbon fiber had better interfacial adhesion than the T700S carbon fiber. These IFSS results of

CF tow and polymeric resins were entirely consistent with the wetting test results thereby indicating a proportionate relation between the electrical resistance results and IFSS.

The proportionate relationship between IFSS and electrical resistance change is further supported by the close relationship between the two quantities shown in Figure 8. The relationship between IFSS and electrical resistance change was conjectured for different CF tow and polymer resins, and indeed as shown in Figure 8 both of their trends very closely follow the same quadratic equation line.

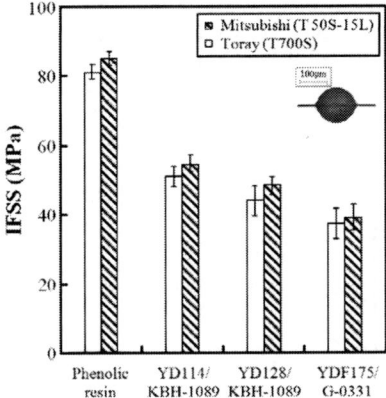

Figure 7. IFSS results of different carbon fibers and polymer resins

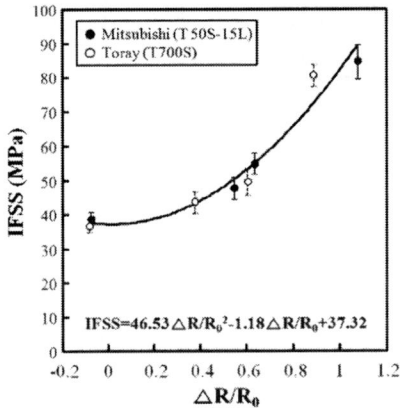

Figure 8. The relationship of IFSS and electric resistance change

Figure 9 shows the ILSS results of CF tow composites with different polymeric resins, and these results very closely duplicate the IFSS results. The relationship and trends between IFSS and the relative changes in resistance shown in Figure 10 further reinforce the conclusion

43

that they are closely related as discussed for figure 8. The rather small differences in the numerical values for ILSS in Figures 9 and 10 and those for IFSS in Figures 7 and 8 are likely due to the specimen types and evaluations, the ASTM D-2344 short beam specimen in the first case and the microdroplet specimen in the second case. In fact, it is stated in the Significance and Use part of ASTM D-2344 that, "In most cases, because of the complexity of internal stresses and the variety of failure modes that can occur in this specimen, it is not generally possible to relate the short-beam strength to any one material property. However, failures are normally dominated by resin and interlaminar properties." Considering this difference the authors find the close agreement between the two test methods very satisfying. For the wetting process in Fig. 8 and 10, the immersion speed of polymer matrix on carbon fiber was different with elapsed time due to the limited matrix amount. Therefore, the trends of IFSS and ILSS were quadratic equation relation.

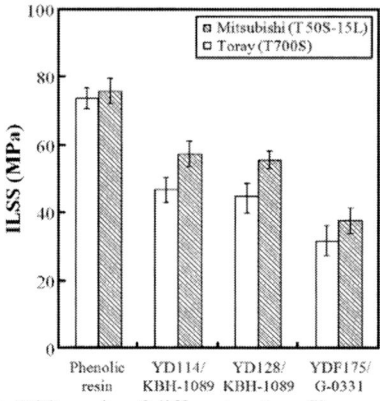

Figure 9. IFSS results of different carbon fibers and polymer resins

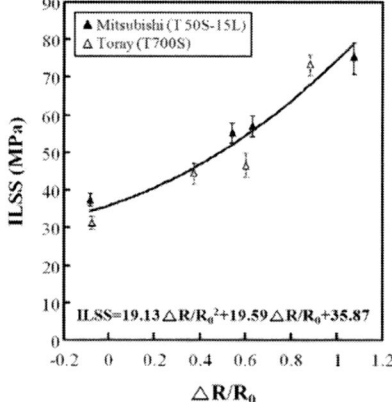

$$ILSS = 19.13 \triangle R/R_0^2 + 19.59 \triangle R/R_0 + 35.87$$

Figure 10. The relationship of ILSS and electric resistance change

CONCLUSIONS

In the research reported here a number studies, all of which were that were thought to be closely related to interfacial properties, were conducted on carbon filled composites with different polymeric matrices. The difference of surface morphology and chemical component of carbon fibers was investigated by FE-SEM and FT-IR tests. The chemical components of two carbon fibers were similar, whereas Toray fibers had more clear ridges and striations parallel on fiber surface. It was supposed that Mishubishi fiber had better wettability with resins due to the fewer ridges. These included: electrical resistance changes during resin wetting, static contact angle measurement to determine the wetting of the carbon tow by the resin materials, IFSS measurements by the microdroplet pull-out tests, and IFSS determined by the ASTM D2344 Short Beam Shear Test. It was concluded that these were all interrelated properties all of which are dependent on interfacial phenomena. Improvement in one resulted in improvement of the others. For example, phenolic resin wet the carbon fiber tows better than any of the epoxy resins studied. It also exhibited the largest change in electrical resistance in the 20 minutes wetting test, as well as the largest IFSS and ILSS according to the microdroplet pullout and short beam shear tests. On the other hand, CF tow/epoxy exhibited the poorest carbon tow wettability, the smallest change in resistance and the poorest IFSS and ILSS. For the four systems studied, the relative change in each of these quantities for the materials of a given composite was roughly proportional to the other measured quantities. This may have some very significant practical importance. It may point the way to use quick easy tests to screen and predict other behaviors for composites of different materials and/or processes. Perhaps a simple resistance change test might be used to predict and select good candidate materials, surface processing techniques for high strength composite applications. The proportional relationship between interfacial adhesion and electrical resistance change was obtained by trend fitting line analyses. Ultimately, it was demonstrated that mechanical properties related to interfacial properties might potentially be predicted by electrical resistance measurement and studies of wetting behavior, using empirical formulas and correlations..

ACKNOWLEDGMENTS

This work was supported by the Agency for Defense Development (ADD) via Korean Institute of Materials Science (KIMM) under contract UE135026GD, 2013.

REFERENCES

1. H. P. Maheshwari, R. B. Mathur, *Electrochim. Acta* 54, 7476 (2013).
2. S. Wang, D. D. L. Chung, *Carbon* 54, 2739 (2006).
3. J. M. Park, J. H. Jang, Z. J. Wang, D. J. Kwon, K. L. DeVries, *Compos. Part B-Eng.* 41, 1702 (2010).

4. X. Jia, G. Li, B. Liu, Y. Luo, G. Yang, X. Yang, *Compos. Part A- App. Sci. Manuf.* 48, 101 (2013).
5. M. L. Sham, J. K. Kim, *Carbon* 44, 768 (2006).
6. R. Zhang, X. Wang, K. K. Shiu, *J Colloid Interface Sci.* 316, 517 (2007).
7. Y. Geng, M. Y. Liu, J. Li, X. M. Shi, J. K. Kim, *Compos. Part A- App. Sci. Manuf.* 1876 (2008).
8. M. Li, Y. Gu, Y. Liu, Y. Li, Z. Zhang, *Carbon* 52, 109 (2013).
9. A. Awal, G. Cescutti, S. B. Ghosh, J. Müssig, *Compos. Part A- App. Sci. Manuf.* 42, 50 (2011).
10. Y. J. Kim, M. Hossain, Y. Chi, *Cold Reg. Sci. Technol.* 67, 37 (2011).
11. J. M. Park, J. H. Kim, *J Colloid Interface Sci.* 168, 103 (1994).
12. R. Tadmor, K. G. Pepper, *Langmuir* 24, 3185 (2008).
13. R. Kannan, V. Vaikuntanathan, D. Sivakumar, *Colloid Surface A* 386, 36 (2011).
14. T. H. Davies, *Mechanism and Machine Theory* 16(3), 171 (1981).
15. Z. J. Wang, D. J. Kwon, G. Y. Gu, H. S. Kim, D. S. Kim, C. S. Lee, K. L. DeVries, J. M. Park, *Composites Sci. Technol.* 81, 69 (2013).
16. J. M. Park, Z. J. Wang, D. J. Kwon, G. Y. Gu, K. L. DeVries, *Solid State Electron.* 79, 147 (2013).
17. C. Li, T. W. Chou, *Composites Sci. Technol.* 68, 3373 (2008).
18. Y. Yang, C. X. Lu, X. L. Su, G. P. Wu, X. K. Wang, *Mater. Lett.* 61, 3601 (2007).
19. J. M. Park, Z. J. Wang, D. J. Kwon, G. Y. Gu, W. I. Lee, J. K. Park, K. L. DeVries, *Compos. Part B-Eng.* 43, 2272 (2012).

Mater. Res. Soc. Symp. Proc. Vol. 1700 © 2014 Materials Research Society
DOI: 10.1557/opl.2014.552

Self-assembled Carbon Nanotube-DNA Hybrids at the Nanoscale: Morphological and Conductive Properties Probed by Atomic Force Microscopy

M. Gabriella Santonicola[1,2], Susanna Laurenzi[3], Peter M. Schön[2]

[1]Department of Chemical Materials Environmental Engineering, Sapienza University of Rome, Via del Castro Laurenziano 7, 00161 Rome, Italy
[2]Materials Science and Technology of Polymers, MESA+ Institute for Nanotechnology, University of Twente, 7500 AE Enschede, The Netherlands
[3]Department of Astronautic Electrical and Energy Engineering, Sapienza University of Rome, Via Salaria 851-881, 00138 Rome, Italy

Corresponding author: mariagabriella.santonicola@uniroma1.it

ABSTRACT

Our research is focused on the engineering of novel, highly sensitive and miniaturized hybrid materials from carbon nanotubes (CNTs) and DNA molecules for applications in biosensors and medical devices. These hybrid sensors allow for a high degree of miniaturization, a key factor in the design of lightweight components while maintaining the advantages of in-situ and real-time analysis capabilities. In the first phase of the sensor design process, we investigated the structural and electrical properties of the supramolecular complexes made of amide-functionalized CNTs and double-stranded DNA. The solubilization properties of the hybrid nanotubes in aqueous solutions with different concentrations of DNA were studied, and an optimal ratio of nanotubes and biomolecules to achieve a good level of dispersion was found. Complexes formed in aqueous solution from CNTs and DNA are highly stable and maintain their properties up to one month from preparation. The morphology of the CNT-DNA composites was investigated at the nanoscale level using atomic force microscopy (AFM) and electron microscopy (SEM). Results from these experiments show the strong affinity between the surface of the amide-functionalized CNTs and the DNA strands. Further, the CNT-DNA films were investigated by atomic force microscopy in the PeakForce TUNA mode to assess the suitability of this technique in determining the local conductive properties of the hybrid films.

INTRODUCTION

Hybrid materials assembled from carbon materials, such as carbon nanotubes (CNTs) or graphene, and biological molecules have emerged in the last decade as a promising class of functional materials for a wide range of applications in biotechnology that span from biosensing to biomedical materials with specific conductive properties. In particular, DNA-functionalized carbon nanotubes have been used for biosensing of ions, glucose or peroxides, owing to the enhanced dispersion of the hybrid structure in aqueous solution [1]. Materials containing such hybrid components show exceptional selectivity and sensing capacity, and have been used as chemical sensors detecting odors with rapid responses and fast recovery times on the scale of seconds [2].

The DNA-assisted dispersion of carbon nanotubes (single-walled) was first demonstrated in solutions containing single-stranded DNA (ssDNA) and used for separation of the nanotubes into fractions [3]. The interaction of carbon nanotubes with biomolecules, including DNA, occurs mainly through non-covalent hydrophobic interactions. These forces are the main driving forces for the assembly of the DNA non polar groups on the hydrophobic wall of the CNTs [4]. However, the exact morphology of the hybrid complex is dictated by several factors and can deviate from a helical wrapping around the nanotube [5]. Most important, to our knowledge, how the non-covalent wrapping affects the electrical properties of the CNTs on the nanoscale has not been investigated. In comparison with other dispersing agents, such as surfactants and polymers, DNA molecules appear to have a better control on the solubilization of the CNTs. In fact, surfactant solutions may present complex microstructure with liquid-liquid phase transitions [6], which can adversely affect the dispersion of CNTs.

The interaction of CNTs and DNA molecules has been investigated by atomic force microscopy (AFM), in particular for single-stranded DNA filaments. Recently PeakForce tapping AFM has been introduced as an AFM-based mapping technique of the mechanical properties of composite films [7]. This technique has been further extended with simultaneous tunneling current measurement capability, in the so called PeakForce TUNA configuration that allows the mapping of electrical properties of conductive surfaces [8].

In this work, we present an investigation by AFM of the morphology at nanoscale level of the assembly formed by multiwalled CNTs and double stranded DNA. The nanoscale characterization is further extended to the mapping of the electrical properties of the hybrid conductive films using the AFM PeakForce TUNA configuration.

EXPERIMENT

Materials

Double-stranded DNA was purchased from Sigma-Aldrich and used as received. Multiwalled CNTs (MWCNTs) functionalized with amide groups were obtained from NanoLab Inc. Solutions containing different ratios by weight of CNTs and DNA were prepared using deionized water and sonicated for 1 h in a cold bath to prevent DNA denaturation.

Morphology characterization

The morphology of the supramolecular complexes formed by CNTs and DNA was investigated by AFM using a NanoScope III multimode setup (Bruker AXS) operated in tapping mode in air. AFM cantilevers were Pointprobe Plus tapping mode silicon probes (Agilent Technologies). Image analysis of CNT-DNA was performed using NanoScope III controlled. Samples for AFM imaging were prepared by depositing the CNT-DNA solution on freshly activated silicon substrates. After incubation for few minutes, samples were rinsed with several drops of water, blotted and dried. Films of CNT-DNA solution after evaporation were characterized by scanning electron microscopy (SEM) using a Tescan Vega LSH instrument operated at 20 kV. SEM images were taken at lower (5 kX) and higher (50 kX) magnification.

Conductivity measurements

The characterization of the conductive properties of the hybrid films on the nanoscale was performed in air using AFM in the PeakForce TUNA configuration. DNA/CNT samples were prepared on gold substrates by drop-casting and dried overnight at 50 °C. Silver glue painting was utilized to provide electrical contact from the gold substrate edge to the underlying metal substrate. A Multimode 8 AFM instrument equipped with a NanoScope V controller and a PF-TUNA module (Bruker AXS) was used. Image processing and data analysis were performed with NanoScope software version 8 and NanoScope Analysis software version 1.40. Pt-Ir coated Si tips on Si cantilevers (SCM-PIT, Bruker AXS) were used in the experiments.

DISCUSSION

Hybrid nanostructures of amine-functionalized multiwalled CNTs and dsDNA in aqueous solutions were prepared by ultrasonication as previously reported [5]. Different ratios of nanotubes and DNA (by weight) were prepared from stock solutions of DNA, and dispersions were sonicated for 1 h in cold bath. After preparation, CNT/DNA solutions showed a significant difference as a function of the amount of DNA used. Optimal dispersions were obtained when equal weights of CNTs and DNA were used. This dispersion was stable even after 2 weeks from preparation (Figure 1).

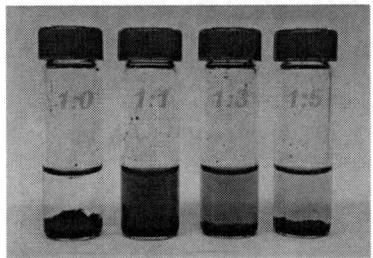

Figure 1. Water solubility of amide-functionalized multiwalled CNTs in DNA (double strand) solutions at different ratios of CNT and DNA (w/w). The photograph was taken 16 days after suspensions were prepared by ultrasonication.

These results that highlight the effect of DNA concentration on the stability of the CNT solubilization was confirmed in SEM investigations. Here the homogeneity of the composite films was analyzed by imaging evaporated films on aluminum supports at lower (5 kX) and higher (50 kX) magnification. Low magnification SEM images showing the morphology of the CNT/DNA films on the micron scale are in Figure 2 and indicate that CNTs are homogeneously dispersed in DNA solutions at 1:1 ratio (by weight), whereas 1:3 and 1:5 preparations lead to the presence of large CNT aggregates in the dispersion. Figure 3 is a high magnification SEM image of CNT/DNA films at 1:1 ratio, where the hybrid nanostructure made of self-assembled DNA around the nanotubes can be identified (red squares).

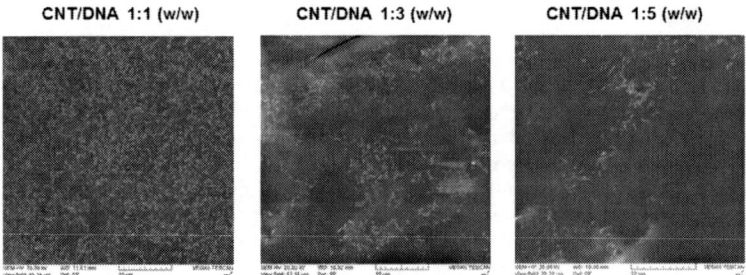

CNT/DNA 1:1 (w/w)   CNT/DNA 1:3 (w/w)   CNT/DNA 1:5 (w/w)

Figure 2. Low magnification SEM images of CNT/DNA films at different ratios of CNTs and double-stranded DNA (w/w). The effect of increasing the DNA concentration on the homogeneity of the composite films is examined. SEM images taken at 5 kX magnification.

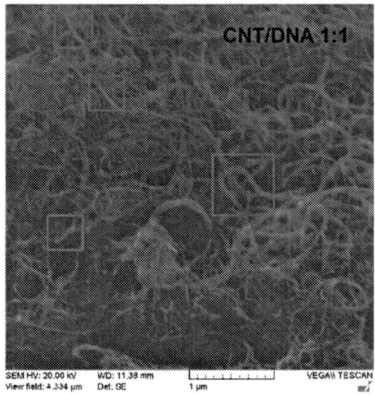

Figure 3. High magnification SEM images of CNT/DNA film at ratio of 1:1 (w/w). Red squares indicate selected hybrid carbon-DNA nanostructures. SEM image taken at 50 kX magnification.

The interaction of amide-functionalized CNTs and double-stranded DNA at the nanoscale level was investigated in AFM experiments (Figure 4). CNTs and DNA-solubilized CNTs were immobilized on a silicon surface and dried. CNTs samples without addition of DNA show highly aggregated features (not shown). The diameter of the CNTs was determined from step-height analysis, and found to be in the range 12-15 nm in agreement with specifications by the manufacturer. When samples of CNTs solubilized by DNA were investigated it was found that DNA wraps along the nanotubes structure forming a characteristic pearl-necklace configuration (Figure 4, left panel). The thickness of the DNA aggregate surrounding the CNTs was determined to be in the range 4-5 nm.

Finally, we investigated the electrical properties across a CNT/DNA film (1:1 w/w ratio) deposited on a gold surface with the AFM in PeakForce TUNA configuration. The right panel of Figure 4 shows the conductive-AFM map obtained by scanning an area of 330 x 350 nm^2 of the hybrid film surface while setting a bias voltage of 200 mV. A current intensity with maximum values around 500 nA was successfully measured. These observations show that conductive-AFM can be effectively used to characterize the electrical properties of the hybrid CNT/DNA structures on the nanoscale. The conductivity of the hybrid film is an important prerequisite in the sensor design, specifically for the detection of UV negative effects on the DNA molecules. In fact, exposure to UV radiation, in particular type C, damages the DNA structure irreversibly and, as a consequence, the conductivity of the hybrid CNT/DNA films will be affected. We plan to extend our preliminary results to the electrical testing of the films in environments characterized by highly damaging UV radiation, including space environment.

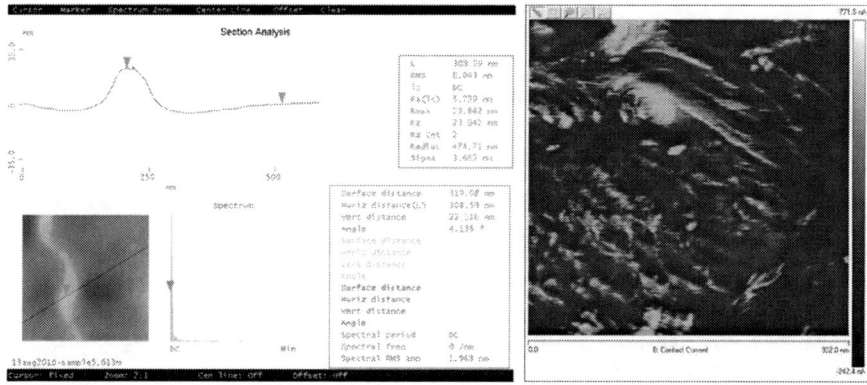

Figure 4. Left panel: high resolution AFM image and section analysis of amide-functionalized multiwalled CTNs wrapped by dsDNA. AFM imaging performed by tapping mode in air with image scan size of 1μm. Right panel: conductive-AFM image of CNT/DNA film (1:1 w/w) showing the contact current measured at bias voltage of 200 mV. Scan size ~ 330 nm.

CONCLUSIONS

In summary we have described the morphological and conductive properties of composite films containing carbon nanotubes that are non-covalently functionalized with double stranded DNA filaments. In this context, the amphiphilic DNA molecule interacts with the CNT walls providing a significant improvement of the dispersion of the hybrid structure in polar solvents, yet the conductive properties of the films on the nanoscale are preserved. The non-covalent functionalization with DNA biomolecules, in contrast with traditional chemical approaches that require oxidation treatments, does not damage the CNT structure thus preserving the electrical properties of the hybrid films.

The results presented in this work are relevant for the engineering of novel hybrid composite materials that can be applied successfully in nanoscale conductive devices for use as biosensors, in particular related to the detection of UV damaging effects on human genome, or bioelectronics suitable in neuronal cell cultures.

ACKNOWLEDGMENTS

We thank Ramón Bueno Morles (Sapienza University of Rome) for the acquisition of SEM images. This work was financially supported by the Italian Ministry of Education, University and Research (MIUR) through the Montalcini Grant to M.G.S.

REFERENCES

1. G. Sanchez-Pomales, L. Santiago-Rodriguez, and C. R. Cabrera, *J. Nanosci. Nanotechnol.* **9**, 2175 (2009).
2. C. Staii, A. T. Johnson, M. Chen, and A. Gelperin, *Nano Lett.* **5**, 1774 (2005).
3. M. Zheng, A. Jagota, E. D. Semke, B. A. Diner, R. S. McLean, S. R. Lustig, R. E. Richardson, and N. G. Tassi, *Nat. Mater.* **2**, 338 (2003).
4. X. Tu and M. Zheng, *Nano Res.* **1**, 185 (2008).
5. M. G. Santonicola, S. Laurenzi, and M. Marchetti, *Proceedings of the 61st International Astronautical Conference 2010,* **13**, 10938-10940 (2010).
6. G. M. Santonicola and E. W. Kaler, *Langmuir* **21**, 9955 (2005).
7. P. Schön, K. Bagdi, K. Molnar, P. Markus, B. Pukanszky, and G. J. Vancso, *Eur. Polym. J.* **47**, 692 (2011).
8. S. Desbief, N. Hergue, O. Douheret, M. Surin, P. Dubois, Y. Geerts, R. Lazzaroni, and P. Leclere, *Nanoscale* **4**, 2705 (2012).

Mechanical Investigation

Mater. Res. Soc. Symp. Proc. Vol. 1700 © 2014 Materials Research Society
DOI: 10.1557/opl.2014.536

Thermo-Mechanical and ILSS Properties of Woven Carbon/Epoxy-XD-CNT Nanophased Composites

Mohammad K. Hossain[1*], Md Mahmudur R. Chowdhury[1], Mahmud B. Salam[1], Johnathan Malone[1], Mahesh V. Hosur[2], Shaik Jeelani[2] and Nydeia W. Bolden[3]

[1]Department of Mechanical Engineering and [2]Department of Materials Science and Engineering, Tuskegee University, Tuskegee, AL 36088, [1*]Corresponding Author: Assistant Professor, hossainm@mytu.tuskegee.edu.
[3]Air Force Research Laboratory Munitions Directorate, Eglin AFB, FL 32542.

ABSTRACT

Carbon fiber-reinforced epoxy composites (CFEC) were fabricated infusing 0, 0.15, 0.30, and 0.40 wt% amino-functionalized XD-grade carbon nanotubes (NH_2-XDCNTs) using the compression molding process under 16 kips. The thermo-mechanical and interlaminar shear properties of CNT incorporated carbon/epoxy composite samples were evaluated by performing dynamic-mechanical thermal analysis (DMTA) and short beam shear (SBS) tests. XD-CNTs were infused into Epon 862 resin using a mechanical stirrer followed by a high intensity ultrasonic liquid processor for better dispersion. After the sonication, the mixture was placed in a three roll milling processor for 3 successive cycles at 140 rpm, with the gap spaces incrementally reduced from 20 to 5 μm, to obtain the uniform dispersion of CNTs throughout the resin. Epikure W curing agent was then added to the modified resin and mixed using a high-speed mechanical stirrer. Finally, the fiber was reinforced with that modified resin using the compression molding process. The results obtained from the DMTA test were analyzed based on the storage modulus, glass transition temperature, and loss modulus. The analysis indicated that the thermo-mechanical properties were linearly increasing from 0 to 0.3 wt% XDCNT loading. The SBS test results exhibited that the incorporation of XDCNTs into the composite increased the interlaminar shear strength (ILSS) by up to 22% at 0.3 wt% CNT loading. Better dispersion of XDCNTs might be attributed to more crosslinking sites and better interaction between fiber and matrix resulting in an improved fiber-matrix interface, whereas, the reaction between functional groups $-NH_2$ of XDCNTs with epoxide groups of resin and epoxy silanes of fiber surfaces improved the crosslinking and thereby ILSS properties of carbon/epoxy composites.

INTRODUCTION

In a fiber reinforced composite, the matrix is the first to fail upon loading because it is the weakest constituent. Hence, the improvement of matrix properties is expected to enhance the overall performance of composites. In the last two decades, researchers have successfully enhanced the matrix properties by incorporating various nanoparticles into epoxy resin and its fiber-reinforced composites.[1-3] CNTs have been proven to be a potential candidate for matrix modification because of its exceptional strength and stiffness, high specific surface area, and high aspect ratio. The higher specific surface area of CNTs facilitates a strong interface for better stress transfer from the matrix to the fiber by bridging effect. The electrical XD-grade CNTs consist of single-walled carbon nanonubes (SWCNT) and double-walled carbon nanotubes (DWCNTs) along with carbon black.[4] XDCNT ($50/g) which is produced by a high yielding process is cheaper than the multi-walled carbon nanotubes (MWCNT) and SWCNT. However, the difference in mechanical properties of nanocomposites consisting of these nanofillers is studied rarely. Okoro et al.[5] studied effects of XDCNTs on the mechanical properties of Epon 862 nanocomposite. They reported improved flexural properties due to the addition of CNTs in the resin matrix while the storage modulus remained unchanged in the glassy region.

Chemical functionalization of the CNT surface was found to further improve the interfacial interaction between CNTs and matrix as well as dispersion of CNTs into the matrix.[6-8] It was reported that covalent bond between amino-functionalized CNTs and epoxy improve the efficiency of load transfer from matrix to fillers resulting in an increase in loss modulus due to more energy dissipation in composites.[9] It was found that the combination of high speed shear mixing and sonication is the best dispersion technique for infusing MWCNTs (0.1-0.4 wt%) in Epon 862 resin with improved thermo-mechanical properties.[3] An increase in dynamic mechanical properties up to 0.3 wt% loading of CNTs with the improvement of around 90% in storage modulus and 22 °C in glass transition temperature (T_g) compared to the reference one was also reported.[3] The effect of amino functionalized MWCNTs on the thermal and mechanical properties of E-glass/epoxy laminated composite cured in the compression hot press was studied.[10] In the last two decades, many researchers have worked to improve the ILSS of fiber reinforced polymer composites (FRPC) by incorporating nanoparticles.[11-15] Wichmann et al.[16] also found 16% enhancement in ILSS properties at 0.3 wt% of functionalized DWCNTs loading. The incorporation of 0.25 wt% of MWCNTs in carbon fiber reinforced epoxy composites manufactured by the vacuum assisted resin transfer molding method enhanced ILSS by 27%.[17] In this study, amino functionalized XDCNTs were dispersed with a combination of sonication and the 3 roll mixing process. Finally, conventional and nanophased carbon fiber/epoxy composites were manufactured using the compression molding process. SBS, DMTA and TMA tests were performed to investigate the effect of XDCNTs on the mechanical and thermal properties of carbon fiber/epoxy composites. Further, fracture morphology studied by scanning electron microscopy (SEM) revealed better interfacial bonding in the CNT-loaded CFEC.

EXPERIMENTAL DETAILS

The matrix used in this study is a two part system. Part A is Epon 862 (Diglycidyl Ether of Bisphenol F), and Part B is Epicure W, an aromatic diamine used as a curing agent of epoxy resin, obtained from Miller Stephenson Chemical Company, Danbury, CT. 8-harness satin weave T300 carbon fabric purchased from US composites with an average thickness of 0.45 mm. XD-grade CNTs functionalized with amine groups (NH_2-XDCNTs) was received from Unidym, Houston. These CNTs had average diameter of 9.5 nm and average length of less than one micron. Concentration of functional groups was less than 4%.

At first, a pre-calculated amount of amino functionalized XDCNTs (0.15, 0.3 or 0.4 wt%) were mechanically mixed with epoxy resin Part-A. The mixture was then put into a sonicator for 1 hour at 35% amplitude and 40s on/20s off cycle pulse mode. The sonicated mixture was then passed through the three rollers to further improve the dispersion of XD-CNTs. In this process, CNTs are further de-agglomerated and uniformly dispersed in resin by the induction of a high shear force in the mixture. The XDCNT/epoxy mixture was then mixed with the curing agent Epikure W using a mechanical mixer according to the stoichometric ratio (Part A: Part B = 100:26.4). Both conventional and nanophased carbon fiber-reinforced epoxy composites were fabricated by employing a combination of hand lay-up and compression hot press techniques.

The apparent interlaminar shear strength (ILSS) of fabricated CFEC was determined using short beam shear (SBS) test as per the ASTM D-2344-00 standard. Six samples for each category were tested using the Zwick-Roell Z 2.5 testing unit. The breaking load P_b noted at this point was then used to calculate ILSS using the relation derived from classical beam theory:

$$ILSS = \frac{0.75P_b}{bXh} \qquad (1)$$

DMA was performed with a TA Instruments dynamic mechanical analyzer (Model Q800) according to the ASTM D4065-01 standard under a dual cantilever beam mode with a frequency of 1 Hz and amplitude of 15 μm. The temperature was ramped from 30°C to 250 °C at a rate of 5 °C/min. TMA tests were carried out on a TA instruments thermo mechanical analyzer (Model Q400) operating in the expansion mode at a heating rate of 5 °C/min from 30°C to 250°C. Five samples of each type were tested. Fracture morphological properties of composite samples were evaluated through SEM study. SEM analysis was carried out using a Zeiss EVO 50.

RESULTS AND DISCUSSIONS

From Table 1, it can be observed that with increase of XDCNT loading, the void content also increases. It can be attributed mostly due to the increase of resin viscosity. The increased viscosity may obstruct the evacuation of entrapped bubbles and volatile impurities during epoxy resin processing. The presence of voids has a detrimental effect on the mechanical properties of CFEC. On the other hand, addition of well dispersed CNTs improves the properties due to better interfacial interactions between modified epoxy and carbon fibers. Observed results in the current study are the net effects of these two opposing phenomena.

SBS test was performed on neat and nanophased CFEC to obtain their ILSS which is a measurement of fiber-matrix binding strength. Generally in SBS test, failure occurs due to a combination of interlaminar shear cracking, micro-buckling, and fiber rupture of the specimens.[18] Figure 1(a) shows the load vs. deflection curves of conventional and 0.15-0.4 wt% carbon/epoxy samples obtained from the SBS test. The ILSS was observed to be the highest in case of 0.3 wt% samples with 22% increase in comparison to control samples. Improved load transferring ability between epoxy matrix and nanotubes and more robust interfacial bonding between matrix and fiber are the conspicuous reasons that are responsible for enhancements in ILSS properties. Generally after mixing epoxy Part A and XDCNT–NH$_2$, the interfacial reaction takes place between amine functional groups of CNTs and epoxide groups of DGEBPF resin, which consists of ring opening reactions followed by a cross-linking reaction.[19] The crosslinking between amino groups of CNTs and epoxide group of resin might result in improved ILSS in 0.3 wt% loading through the possibility of effective stress transfer between epoxy and CNTs. When compared to the 0.3 wt% sample, the ILSS of 0.4 wt% sample was decreasing. This decreasing might be attributed to the formation of excessive agglomeration. At higher loading, CNTs get closer to each other in the matrix; hence their strong attractive forces might have led them to form agglomerates. These agglomerates can act as stress risers and also increase the free volume by creating voids in the matrix.[20] The fractured surface of nanophased CFEC demonstrates the presence of resin sticking to the fiber surface (Fig. 2b), which is an indication of stronger interfacial bonding between the fiber and matrix. On the other hand, for the conventional CFEC shown in Figure 2(a), the fiber surface appears quite smooth and very little amount of resin adheres to the fiber. Moreover, the resin does not show any sign of protrusion on fiber surface (Figure 2a). These phenomena indicate a weak interfacial bonding between the fiber and matrix.

The coefficient of thermal expansion (CTE) is an important thermo-mechanical property of polymeric composites for engineering applications. Most of the polymeric materials have high CTE value which limits their applications. However, an incorporation of a small amount of

Table I. Physical properties of carbon/XDCNT/epoxy composites.

Properties	Sample Specification			
	Neat	0.15 wt%	0.3 wt%	0.4 wt%
Fiber Volume Fraction (%)	71.56±0.25	70.95±0.66	71.22±1.17	72.19±0.41
Void Fraction (%)	1.87±0.39	2.27±0.75	2.13±0.48	2.45±0.14
Density (g/cm^3)	1.592±0.016	1.582±0.022	1.584±0.035	1.589±0.018
ILSS (MPa)	36.39±2.18	37.37±1.92	44.32±1.35	41.43±2.35

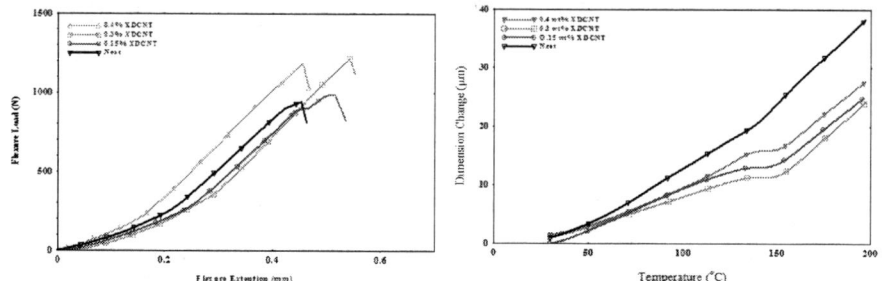

Figure 1. a) Load vs. deflection curve & b) Dimension change with temperature for CFEC.

nanoparticles as filler material in the matrix can significantly reduce the overall CTE of carbon/epoxy composites.[21] A low CTE value is most desirable to ensure good dimensional stability. Figure 1b presents the change in sample dimension in transverse direction as a function of temperature. The average CTE of the control sample was found to be 51.87 ppm/°C. The addition of 0.3 wt% nanotubes reduced this value to 32.4 ppm/°C which represents a maximum decrease of 37.54% (Table 2). The improved dispersion of CNTs in matrix facilitates to enhance interfacial reactions and forms covalent bond between them. This covalent bond induces different crosslinking regions into the epoxy matrix. Two ring opening reactions followed by a cross-linking reaction create interlocking structure in the resin blend those hindrances the mobility of polymer chains in the matrix system.[3] Thus, the reduction of CTE after addition of CNTs was due to its improved dispersion in the epoxy and the reduced segmental motion of the epoxy matrix.[22] Moreover, well dispersed CNTs can align the polymer chain along their axial direction, thus can easily get associated with the polymer molecule and prohibit its thermally induced movement resulting in a reduced CTE value.[23] The addition of 0.4 wt% CNTs resulted in slight increase in the CTE as compared to that of the 0.3 wt% sample but still low in comparison with control system. This higher value of CTE can be demonstrated by the aggregates formed at a higher loading as explained in earlier sections.

Figure 3 shows the temperature dependence of storage modulus in the range of 30-250 °C. For different laminates, storage modulus at 30°C has been presented in Table 2. It can be seen that the addition of CNT concentration has resulted in significant improvement in storage modulus. Specifically, the storage modulus was improved by 17.22% with the addition of 0.3 wt% of XDCNT. This improvement is due to enhanced interaction between well dispersed CNTs and the matrix. More sites for XDCNT/polymer interaction have been provided by better dispersion of XDCNTs at 0.3 wt%. The formation of strong covalent bonds attributed to the presence of amino functional groups of XDCNT and its reaction with epoxy. The epoxy chain molecular motion around nanotubes has been abridged by the formation of covalent bond and the

Figure 2. Interfacial bonding in (a) neat sample (X2000) and b) 0.3 wt% sample (X2000).

enhanced interaction. This abridgement may have resulted in a significant change of elastic and viscous properties in nanocomposites. However, at 0.4 wt% loading, the observed decrease instorage modulus can be attributed to increase of agglomeration of XDCNTs which reduces the crosslinking sites. Therefore, the storage modulus decreases as a result of increased molecular motion and movement of the chain. As in storage modulus, loss modulus was also increased up to 0.3 wt% CNTs compared to the control sample. The tanδ vs. temperature relationship in Figure 3 demonstrates the effect of nanotube concentration on damping properties of laminated composite. The addition of small amount of CNTs up to 0.3 wt% is observed to increase T_g slightly. The increase in T_g in the polymeric system is predominantly affected by the amount and dispersion of CNTs, degree of crosslinking, and interfacial interaction.[24] When the CNTs are well dispersed the molecular motions are restricted and degree of crosslinking and interfacial interaction are improved. A positive shift of the tanδ curves is the indication of increase in T_g.

Table II. Thermo-mechanical properties of carbon/XDCNT/epoxy composites.

Specimen Category	Storage Modulus (MPa)	% Change w.r.t control	Loss Modulus (MPa)	% Change w.r.t control	Glass Transition Temperature (°C)	% Change w.r.t control	CTE (μm/(m-°C))	% Change w.r.t control
Neat	18118±249	-	1519±16	-	155.7±1.34	-	51.87±2.53	
0.15 wt%	19308±292	6.57	1558±21	2.57	158.43±1.43	2.73	46.46±1.85	-10.43
0.3 wt%	21239±166	17.22	1736±37	14.31	161.21±2.38	5.51	32.4±1.93	-37.54
0.4 wt%	19275±364	6.38	1576±50	3.75	155.99±0.96	0.29	44.4±2.41	-14.4

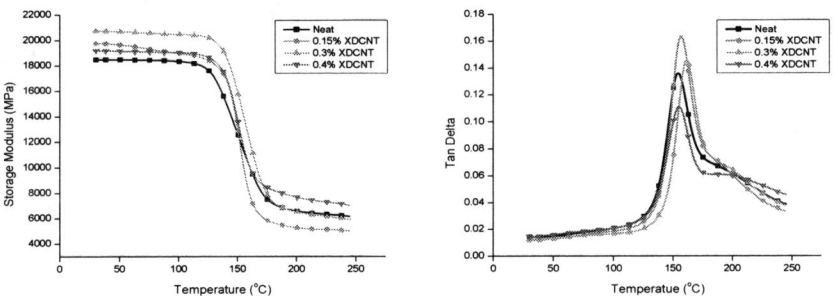

Figure 3. Temperature dependence of storage modulus and loss factor for CFRP composites.

CONCLUSIONS

In this study, amino functionalized XDCNTs were infused as nanofillers in woven CFEC. The incorporation of XDCNTs at very low concentration (up to 0.3 wt%) enhanced the ILSS and thermo-mechanical properties of CFRP significantly. The storage modulus was increased by 17.22% at 0.3 wt% loading in comparison to conventional CFRP composites. The loss modulus and T_g had also been improved slightly with the addition of XDCNTs. Lowest CTE was found in the 0.3 wt% XDCNT-infused CFRP composites. The ILSS was also observed to have the highest value at 0.3 wt% loading due to better interfacial interaction and effective load transfer between NH_2- XDCNTs surfaces promotes good adhesion between carbon fiber and epoxy matrix.

ACKNOWLEDGMENTS

The authors acknowledge the Air Force Research Laboratory Munitions Directorate, Eglin AFB, FL 32542, USA and NSF-EPSCoR for their financial support to carry out this research work.

REFERENCES

1. P. Farhana, Y. Zhou, V. Rangari and S. Jeelani, *Mater. Sci. Eng.: Part A 405*, 246 (2005).
2. H. Mahfuz, A. Adnan, V. Rangari, S. Jeelani and B.Z. Jang, *Compos. Part A 35*, 519 (2004).
3. Y. Zhou, P. Farhana, L. Lewis and S. Jeelani, *Mater Sci. Eng: A 452*, 657 (2007).
4. K. Tao, S. Yang, J. C. Grunlan, Y. S. Kim, Dang, B. Y. Deng, R. L. Thomas, B. L. Wilson, and X. Wei, *J. Appl. Polym. Sci. 102*, 5248 (2006).
5. C.U. Okoro, M.K. Hossain, M.V. Hosur, S. Jeelani, *J. Eng. Mater. Technol.* 133, 131 (2011)
6. Z. Yaping, Z. Aibo, Z. Jiaoxia, and N. Rongchang, *Mater. Sci. Eng. A 145*, 435 (2006).
7. X. Xie, Y. W. Mai, and X. Zhou, *Mater. Sci. Eng. A 49*, 89 (2005).
8. N.G. Sahoo, S. Rana, J. W. Cho, L. Li, *Prog. Polym. Sci. 35*, 837 (2010,).
9. D. P. N. Vlasveld, W. Daud, H. E. N. Bersee, and S. Picken, *Compos. Part A 38*,730 (2007).
10. M. M. Rahman, S. Zainuddin, M. V. Hosur, J. E. Malone, M. B. A. Salam, A. S. Kumar and S. Jeelani, *Compos. Struct. 94*, 2397 (2012).
11. F. Rosselli and M. H. Santare, *Composites, Part A* 28(6), 587 (1997).
12. T. Yokozeki, Y. Iwahori and S. Ishiwata, *Composites, Part A 38*, 917 (2006).
13. N. A. Siddiqui, R. S. C. Woo, J. K. Kim and A. Munir, *Composites, Part A 38*, 449 (2006).
14. J. Zhu, A. Imam, R. Crane, K. Lozano, E.V. Barrera, Compos. Sci. Technol. *67*, 1509 (2006).
15. F. H. Gojny, M. H. G. Wichmann, B. Fiedler, W. Bauhofer and K. Schulte, *Composites, Part A 36*, 1525 (2005).
16. M. H. G. Wichmann, J. Sumfleth, F. H. Gojny, M. Quaresimin, B. Fiedler and K. Schulte, Eng. Fract. Mech. *73*, 2346 (2006).
17. E. Beyakrova, E. T. Thostenson, A. Yu, H. Kim, J. Gao, J. Tang, H. T. Hahn, T. W. Chou, M. E. Itkis and R. C. Haddon, Langmuir, 23, 3970 (2007).
18. A.K. Subramaniyan and C. T. Sun, *Composites, Part A 37*, 2257 (2006).
19. P. C. Ma, S. Y. Mo, B. Z. Tang and J. K. Kim, *Carbon 48*,1824 (2010).
20. A. Y. Zhang, D. H. Li, H. B. Lu, H. Y. Xiao and J. Jia, *Express Polym. Lett. 5*, 708 (2010).
21. M. M. Shokrieh, A. Daneshvar and S. Akbari, *Carbon*, 59, 255(2013).
22. K. P. Pramoda, N. T. T. Linh, P. S. Tang, W. C. Tjiu, S. H. Goh, C. B. He, *Compos. Sci. Technol. 70*, 578 (2010).
23. H. R. Lusti and A. A, Gusev, Model. Simul. Mater. Sci. Eng. 12, 107 (2004).
24. S. Ganguly, A. K. Roy and D. P. Andersion, *Carbon* 46,806 (2008).

Mater. Res. Soc. Symp. Proc. Vol. 1700 © 2014 Materials Research Society
DOI: 10.1557/opl.2014.715

Molecular Dynamics Simulation of a Pullout Test on a Carbon Nanotube in a Polymer Matrix

Guttormur Arnar Ingvason[1] and Virginie Rollin[1]

[1]Embry-Riddle Aeronautical University, Aerospace Engineering Department, 600 South Clyde Morris Blvd, Daytona Beach, FL 32114, U.S.A.

ABSTRACT

Adding single walled carbon nanotubes (SWCNT) to a polymer matrix can improve the delamination properties of the composite. Due to the complexity of polymer molecules and the curing process, few 3-D Molecular Dynamics (MD) simulations of a polymer-SWCNT composite have been run. Our model runs on the Large-scale Atomic/Molecular Massively Parallel Simulator (LAMMPS), with a COMPASS (Condensed phase Optimized Molecular Potential for Atomistic Simulations Studies) potential. This potential includes non-bonded interactions, as well as bonds, angles and dihedrals to create a MD model for a SWCNT and EPON 862/DETDA (Diethyltoluenediamine) polymer matrix. Two simulations were performed in order to test the implementation of the COMPASS parameters. The first one was a tensile test on a SWCNT, leading to a Young's modulus of 1.4 TPa at 300K. The second one was a pull-out test of a SWCNT from an originally uncured EPON 862/DETDA matrix.

INTRODUCTION

Carbon Fiber Reinforced Plastic (CFRP) has been researched and used for many decades but still has the structural drawback of only being resilient in the fiber direction. This can be alleviated using layers with different fiber directions. The thickness strength (perpendicular to the fibers) of the CFRP or any composite material is not nearly as high as in the fiber direction [1]. Crack initiation and growth is a weakness of CFRP. These crack failures often present themselves as delamination at the interface between two layers. In order to improve the strength of CFRP and prevent or delay delamination, Single Walled Carbon Nanotubes (SWCNTs) can be added to the polymer matrix. If a minute delamination occurs in between two layers of composite, the carbon nanotubes (CNTs) act as a barrier and stop the delamination from propagating further between the layers. The CNT bridges the gap in the resin and deflects the crack propagation [1]. This improves the delamination toughness of the composite [2]. The study of CNT/polymer pull out tests is important in understanding how CNTs can assist in delaying or preventing delamination in CFRP. The goal of this research is to implement the Condensed phase Optimized Molecular Potential for Atomistic Simulations Studies (COMPASS) potential parameters in the Large-scale Atomic/Molecular Massively Parallel Simulator (LAMMPS) [3]. The potential is then used in two different simulations to test its implementation. The first is a simple tension test on a SWCNT and the second one is a pull out test with a SWCNT and a mix of EPON 862 resin with a DETDA curing agent. A pullout simulation gives information on the interaction between CNT and polymer matrix, and hence the strength of the composite.

FORCE FIELD

The COMPASS force field was chosen for implementation in the MD simulations. COMPASS is an ideal potential since it can be applied to both the CNT and the polymer curing process. The ReaxFF potential could have been another candidate and might be explored in the future. COMPASS was developed by Huai Sun of Molecular Simulations Inc [4] (now Accelrys). This force field was chosen over others due to the promising results that it had shown in simulations conducted in Material Studio by Accelrys [5]. The drawback of the Materials Studio implementation is that the user has no control over the parameters of the potential. LAMMPS, which is an open source code, offers an implementation of the COMPASS potential which allows us to control our own parameters. The values of the parameters for the force field were all taken from references [4], [6] and [7]. The total energy of the system can be split up into two categories [4]:

1) Valence terms that include diagonal and off-diagonal cross-coupling terms that characterize internal coordinates of bond (b), angle (θ), torsion angle (φ), and out-of-plane angle (χ), and the cross-coupling terms include combinations of two or three internal atoms.

2) Non-bonded interaction terms that contain a LJ-9-6 function for the van der Waals and Coulombic term.

SIMULATIONS DETAILS

Tension test on a SWCNT

After setting up the parameters for the COMPASS force field in LAMMPS a tension test on a (60,0) SWCNT was performed. The periodic SWCNT, composed of 24000 atoms, was 425 Å long and had a diameter of 46.98 Å. The CNT's list of coordinates, bonds, angles and dihedrals were created using the Visual Molecular Dynamics (VMD) software [8]. An energy minimization was performed on the model, after which the temperature of the simulation was set to 300K and the (60,0) SWCNT was subjected to a strain rate of 6.25×10^{-8} /fs. Periodic boundary conditions were applied in the Z direction (longitudinal axis of CNT) and fixed boundary conditions in X and Y directions. The strain rate was applied to all the atoms in the system using a constant engineering shear strain rate.

In the COMPASS force field, a bond can stretch indefinitely without breaking. To remediate this issue, LAMMPS has introduced a "bond/break" command, which deletes a bond once it reaches a specified length. The C-C bond breaking distance was set to 1.771 Å [9].

Pull-out test

The second simulation was that of a SWCNT embedded in an uncured mix of EPON 862/DETDA. The ratio of EPON 862 to DETDA was 2:1. The SWCNT is a (10,10) with a diameter of 13.56Å and a length of 150Å. The model was composed of 2480 atoms for the SWCNT, 23985 atoms for the polymer, for a total of 26465 atoms.

The EPON 862 and DETDA molecules were created using the prebuilt EPON 862/DETDA model from the Nanohub Polymer Modeler [10]. The structure obtained from Polymer Modeler contained 16000 atoms. A single copy of the EPON 862 and DETDA molecules was extracted from that information along with the bonds, angles, dihedrals and impropers. Duplicating of EPON and DETDA molecules around the SWCNT was performed using the Packmol software package [11]. To increase the density of the polymer, a lower density was first achieved in a box with a larger size. The box was then shrunk while a force field (50 kcal/Å) of cylindrical shape was set around the CNT to avoid polymer molecules from entering this area. The shrinking was achieved in multiple steps. After each shrinking a minimization was performed to allow the added energy to be released and the new atom positions to be computed. As recommended in [12] the non-bonded interactions between the atom and its third neighbor are reduced by half. Using this method, an initial polymer density of 0.148 g/cm^3 is increased to 0.28 g/cm^3. The target density of 1.2 g/cm^3 was not achieved because of issues getting the polymer atoms into the simulation box without interfering with each other or the SWCNT that was placed in the center of the simulation box. In future work the process of shrinking the simulation box will be refined by using slower shrinking rates [13] so that a density of 1.2 g/cm^3 can be achieved. The lower polymer density may lead to slower curing rates and less contact between the CNT and polymer during the pullout simulation, which may lead to differences in the pull out force.

During our initial literature review, we noticed that the polymer curing process cannot be simulated using any potential. In [13] an Optimized Potential for Liquid Simulation (OPLS) potential was used during the curing process but ReaxFF was used for the rest of the simulation. OPLS in LAMMPS evaluates the non-bonded interactions intermolecularly and intramolecularly for atom pairs separated by three or more bonds. This leads to the doubling of the potential energy between an atom and its third neighbors [12]. This could be countered in LAMMPS, except that it then prevents us from using the command to create or break bonds, and thus regulate the curing process. It is not possible to do both at the same time and these difficulties led to the use of COMPASS for our simulations.

Cross-linking between the EPON 862 and DETDA molecules occurs when an H_2N bond in DETDA and O-CH_2 bond in EPON 862 are broken. The loose H atom then bonds the O atom in the EPON 862. Finally the HN branch on the DETDA molecule bonds with the CH_2 branch in the EPON 862 [5]. We conducted initial simulations where both the N-H_2 and O-CH_2 bonds were broken by the simulation using a LAMMPS command in order to start the cross-linking process. Since the H in N-H_2 (DETDA) only possesses one bond, when it was broken it was loose in the simulation box. Before it could connect to the oxygen in the EPON 862 molecule the simulation would stop because the hydrogen atom would leave the simulation box. To allow for the bond creation process to be completed in the simulation, the O-CH_2 bond was broken manually before the beginning of the simulation in the EPON 862 while allowing one of the hydrogen atoms in NH_2, DETDA molecule, to bond to the oxygen atom.

The initial models would reach the maximum allowed number of bonds per atom within 1000 fs, when around 100 H-O bonds had been created for only a dozen N-CH_2 bonds. To offset this, the creation of the H-O bonds was artificially slowed down by a factor of 2. This permits a more even number of bond types to be created. The system was then allowed to cross-link at a temperature of 600K. For bonds to be created, both atoms have to be eligible to create a bond. Those are the Hydrogen and Nitrogen atoms in the DETDA molecule and the Oxygen and Carbon atoms in the EPON 862 molecule. The atoms have to come within close proximity of each other. A radius was defined in LAMMPS for which the bond was allowed to be created. For the C-N bond 3.0 Å was used, which is approximately twice the equilibrium bond distance, 2.5 Å was used for the O-H bond, which is roughly 2.5 times the equilibrium bond distance. The curing process was run for 1.5 ps with one time step being 0.01 fs. The curing process is expected to go faster once the density of the polymer has been adjusted to 1.2 g/cm³.

Once the polymer has been allowed to cure and energy to be minimized, the temperature was reset to 300K. Upper and lower layers of the polymer were frozen and a constant velocity of 150 Å/fs was added to the SWCNT to start the pullout simulation (Figure 1). The simulation had periodic boundary conditions in the X and Y directions while the Z direction (longitudinal) had a fixed condition that allowed it to resize as the CNT was pulled out.

a)   b)

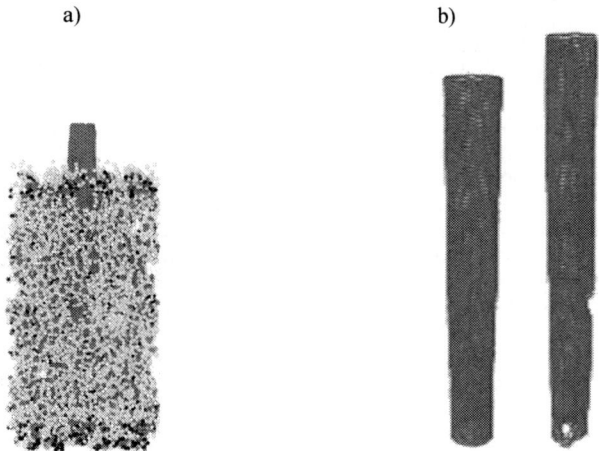

Figure 1. (a) (10,10) SWCNT during pullout test; (b) (60,0) SWCNT before tension test (left) and after tension test (right) note the deformation that can clearly be seen on the bottom half of the tube.

RESULTS

Tensile test

The resulted break of the SWCNT occurred at $\varepsilon = 19.2\%$ and $\sigma = 270$ GPa (Figure 2). These results were obtained using the virial stress of the system [14] and assuming the thickness of the CNT to be 3.4 Å when calculating the volume. This value is of importance since the calculated breaking stress and therefore the obtained Young's modulus of the CNT can vary depending on what value is chosen for the CNT thickness [15]. Assuming the CNT thickness to be 3.35 Å changes the Young's modulus to 1.46TPa. A value of 3.4 Å represents the Van der Waals diameter of the carbon atom [16]. Table I shows a comparison of the results from the current work with that of simulations and experiments from the literature.

$$A = \pi[(r + t/2)^2 - (r - t/2)^2] \qquad (1)$$

Where: A = area used in calculation of Young's modulus, r = radius of CNT and t = 3.4 Å

Table I. Young's moduli form literature as a comparison to current results

Source	CNT (m,n)	Temp. (K)	Young's (TPa)	Method
Current work	(60,0)	300	1.44	MD
[17]	(60,0)	300	0.9-0.92	MD
[18]	(5,5)	300	1.424	MD
[19]		300	1.25-0.35/+0.45	Experimental
[20]	Various	300	0.9298±0.0115	MD
[21]	Various (n,n)	293	1.350±0.012	MD

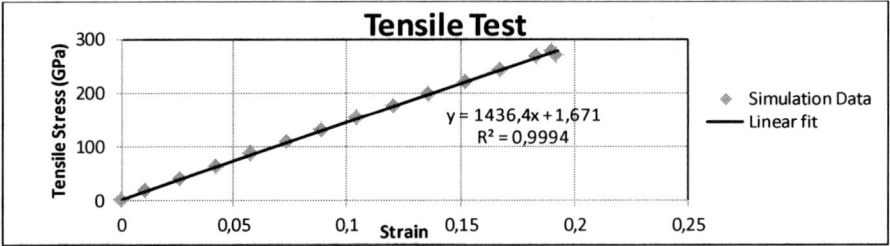

Figure 2. Stress-strain curve of a (60,0) SWCNT

Pull-out test

The initial total energy of the pullout simulation was 130355 kcal/mol, consistent with the literature [5]. In the same reference by Gou et al. the pullout energy is defined as:

$$E_{Pullout} = E_{final} - E_{initial} \qquad (2)$$

The interfacial shear strength is also defined by Gou et al. [5] as:

$$\tau_i = \frac{E_{Pullout}}{\pi * r * L^2} \qquad (3)$$

Where: L^2 = Length of the SWCNT, r = diameter of the SWCNT

In our simulations so far the final total energy of the system rose to more than 1000 times the initial energy. It is currently not possible to calculate the shear stresses since the initial and final energies are needed. With the starting energy close to the initial energy of similar simulations this shows that the issues are with the setup of the simulation (boundary condition, application of pull-out force) and not with the potential.

CONCLUSIONS

COMPASS force field parameters were implemented for use in MD simulations that are performed using LAMMPS. The user has full control over the input of the potential, which, as noted by J. Tack in [20] is an issue when using in Material Studio by Accelrys. This implementation was tested on two different setups, a tensile test of a SWCNT and a pull-out test of a SWCNT from a polymer matrix.

The initial results are promising for the plain (60,0) SWCNT in tension. The calculated Young's modulus at 1.4 TPa is on the higher side of the results from the literature but more simulations with different CNT (m,n) and lengths must be performed. The relatively high Young's modulus may be due to the intramolecular non-bonded interactions not being turned off or weighted. Future work can include simulations where the atom interactions are weighted and the failure is based on bond length calculations that are performed either using LAMMPS commands or a separate code that uses the LAMMPS output to calculate the bond length. The SWCNT embedded in the partially cured EPON 862/DETDA also showed very good early results for the pullout simulation. Current focus on reaching the target density of 1.2 g/cm^3 is showing good progress. Moreover, an adjustment of the boundary conditions is still needed to keep the energy of the system within a physically meaningful range.

ACKNOWLEDGEMENTS

The authors would like to thank Embry-Riddle Aeronautical University for the support during this research.

REFERENCES

1. S. Khan and J.-K. Kim, *International J. Aero. and Space Sci.*, pp. 112-133 (2011).
2. L. Yang, L. Tong and X. He, *Comp. Mat. Sci*, **55**, pp. 356-364 (2012).
3. S. Plimpton, *J. Comp. Phys.*, p. 117 (1995).
4. H. Sun, *J. Phys. Chem. B*, **102**, pp. 7338-7364 (1998).
5. J. Gou et al., *Comp. Mat. Sci.*, **31**, pp. 225-236 (2005).
6. H. Sun, *Spectrochimica Acta Part A*, **23**, pp. 1301-1323 (1997).
7. H. Sun, *Comp. and Theoretical Polymer Sci.*, **8**, pp. 229-246 (1998).
8. W. Humphrey, A. Dalke and K. Schulten, *J. Molec. Graphics*, **14**, pp. 33-38 (1996).
9. W. Duan, Q. Wang, K. Liew and X. He, *Carbon 45*, pp. 1769-1776 (2007).
10. G. Klimeck, et al., IEEE CISE, **10**, pp17-23, (2008).
11. L. Martínez, et al. *J of Comp. Chem.*, **30**, no. 13, pp. 2157-2164 (2009).
12. R. C. Rizzo and W. L. Jorgensen, *J. Am. Chem. Soc.*, **121**, pp. 4827-4836 (1999).
13. G. M. Odegard, et al., *AIAA*, National Harbor, (2014). DOI: 10.2514/6.2014-0467
14. E. T. Lilleoden, et al., *J. Mech. Phys. of Solids*, **51**, pp. 901-920 (2003).
15. Y. Huang, J. Wu and K. Hwang, *Phys. Rev. B*, pp. 245413-1 - 245413-9 (2006).
16. S. S. Batsanov, *Inorganic Mat.*, **37**, pp. 871-885 (2001).
17. J. Gu and F. Sansoz, *MRS Fall Meeting*, Boston (2008). Man. ID: 1137-EE10-05.R1
18. J. J. Oh, M.S. Thesis, Naval Postgraduate School, (2003).
19. A. Krishnan, et al., *Phys. Rev. B*, **58**, no. 20, pp. 14 013-14 019 (1998).
20. B. WenXing, Z. ChanChun, C. WanZhao, *Physica B*, **352**. pp 156-163 (2004).
21. Y. Jin, F.G. Yuan, *Compos. Sci. Technol.*, **63**, pp. 1507-1515 (2003).
22. J. L. Tack, M.S. Thesis, Texas A&M University," (2006).

Optical Investigation

Mater. Res. Soc. Symp. Proc. Vol. 1700 © 2014 Materials Research Society
DOI: 10.1557/opl.2014.574

Coupled Vibrations in Index-Identified Carbon Nanotubes

Dmitry Levshov[1,2], Thierry Michel[1], Matthieu Paillet[1], Xuan Tinh Than[1,3], Huy Nam Tran[1], Raul Arenal[4], Abdelali Rahmani[5], Mourad Boutahir[5], Ahmed-Azmi Zahab[1], Jean-Louis Sauvajol[1]
[1]University Montpellier 2-CNRS, Laboratoire Charles Coulomb, F-34095 Montpellier, France.
[2]Faculty of Physics, Southern Federal University, Rostov-on-Don, Russia.
[3]Laboratory of Carbon Nanomaterials, Institute of Materials Science, VAST, Hanoi, Vietnam.
[4]Laboratorio de Microscopias Avanzadas (LMA), Instituto de Nanociencia de Aragon (INA), University of Zaragoza, Zaragoza, Spain.
[5] Research Team Computational Physics and Nanomaterials, Moulay Ismail University, Meknes, Morocco.

ABSTRACT

Combining high resolution transmission electron spectroscopy, electron diffraction, and resonant Raman spectroscopy experiments on the same suspended (free-standing) individual carbon nanotubes is the ultimate approach to relate unambiguously the structure and the intrinsic phonon features of these nano-systems.
By using this approach, the effect of coupling between nanotubes on the phonons is investigated in two model nano-systems: (i) a bundle of two non-identical SWNTs (inhomogeneous dimer), (ii) double-walled carbon nanotubes.

INTRODUCTION

The combination of high resolution transmission microscopy (HRTEM), electron diffraction (ED) and resonant Raman spectroscopy (RRS) on an individual, spatially isolated, and suspended carbon nanostructure is the ultimate method to relate unambiguously its structural parameters, optical transitions and Raman-active phonon modes. From 2005, our group has followed this approach to determine the radial breathing mode (RBM) and the G-modes features, as well as to evaluate the transition energies of individual, achiral and chiral, semiconducting and metallic, index-identified suspended single-walled carbon nanotubes (SWNTs) [1-5].

In this communication, we apply the same approach on two van der Waals-coupled carbon nanostructures, namely: (i) a bundle of two non-identical SWNTs (inhomogeneous dimer) and (ii) double- walled carbon nanotubes (DWNTs).

An inhomogeneous dimer of SWNTs is characterized by a weak van der Waals interaction between the neighboring carbons of its two constituting SWNTs. Dimers of SWNTs have three electronic configurations, namely: SC-SC, SC-M and M-M, where SC (resp. M) stands for semiconducting (resp. metallic) SWNT.

DWNT is a coaxial structure composed of two layers: an inner and an outer SWNT. DWNTs display four different electronic configurations, namely: SC@SC, SC@M, M@SC and M@M. Properties of DWNTs are related to the individual nature of each layer and to the layer interactions. It was found that the same inner tube can be contained inside different outer tubes [6] leading to different layer interaction strengths with respect to the interlayer distance.

In this paper, vibrations of these two coupled nanostructures are investigated by Raman spectroscopy. We focus here on the low-frequency range of the Raman spectrum where the breathing modes are detected.

EXPERIMENT

The individual carbon nanostructures (dimer of SWNTs and DWNTs) were synthesized by chemical vapor deposition (CVD) directly onto commercial TEM grid with holes (2 μm in diameter).[7]

TEM, HRTEM and electron diffraction patterns (ED) were recorded in a FEI Titan microscope operating at 80 kV to reduce damages induced by electron irradiation. For the same reason, TEM images and EDP were recorded within an average 5 s acquisition times.

Resonant Raman scattering measurements were carried out using a Jobin Yvon T64000 spectrometer equipped with a liquid nitrogen-cooled silicon CCD detector. The scattered light was collected through a microscope using a backscattering configuration. In all the measurements, both incident and scattered light polarizations are along the nanotube axis (// // polarized Raman spectrum). Incident excitations from Ar+ and Kr+ lasers, a Dye laser and a tunable Ti/sapphire laser were used. In order to avoid heating effects, the laser power impinging the sample was kept below 50μW with a 100x objective (Numerical Aperture of 0.95).

RESULTS

Raman spectroscopy of index-identified dimer of single-walled carbon nanotubes

An individual free-standing two-nanotube bundle (dimer) of SWNTs was identified from HRTEM and its structure was determined by ED. This dimer is composed by two interacting semiconducting SWNTs, namely: the (15, 11) and (16, 12) SWNT. These tubes have close but not identical diameters (1.77 nm for (15,11) and 1.91 nm for (16,12)), and their chiral angles are both close to 25°. In figure 1, the low-frequency range of the Raman spectrum excited at 1.92 eV is displayed. To the best of our knowledge, it is a second example of Raman spectrum measured on an individual index-identified inhomogeneous dimer of SWNTs (see Ref. 8 for the first example).

The spectrum of the (15,11)-(16,12) dimer displays four bands (called BM) located at 124.5 cm^{-1} (BM4), 131 cm^{-1} (BM3, the most intense), 138 cm^{-1} (BM2) and 146 cm^{-1}(BM1). The presence of additional modes as compared to the two expected RBM is attributed to the lowering of the symmetry in dimers [9,10]. On the basis of preceding theoretical works [see figure 4, in Ref. 9], it is tempting to assign the lower-frequency pair of peaks (124.5 and 131 cm^{-1}) to vibrations of the (16,12) SWNT and the higher-frequency pair (138 and 146 cm^{-1}) to the vibrations of the (15,11) SWNT. This behavior has been predicted for tubes of small diameter with rather different vibrational frequencies when isolated [10]. However, in the (15,11)-(16,12) dimer, the SWNTs diameters are relatively large and close to each other. In addition, the frequencies of their respective radial breathing modes are relatively close when isolated (expected at 134 cm^{-1} and 142 cm^{-1} respectively from the ω_{RBM} (d) established for SWNTs [5]). In consequence, we propose that each BM peak can be associated to a mixing of breathing modes of different symmetries (RBM and E_n modes of the isolated SWNTs) [8,9,10]. This hypothesis is supported by the fact that only the (16,12) SWNT is resonant at 1.92 eV [3] while all the four modes are observed at this wavelength.

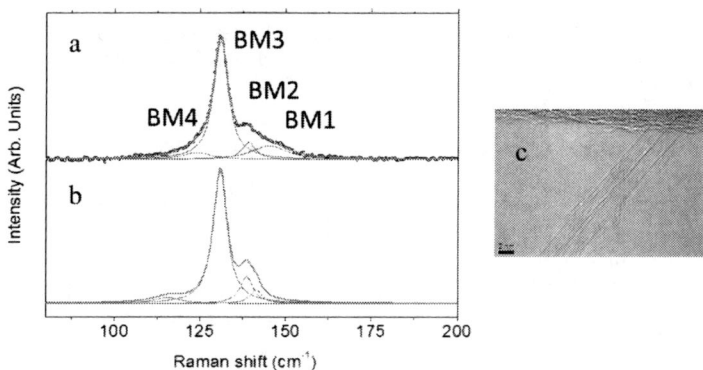

Figure 1: (15,11)-(16,12) dimer: (a) Raman spectrum excited at 1.92 eV, (b) the calculated spectrum (see text), (c) High-resolution micrograph .

Due to the weak van der Waals interaction between the tubes, only a small shift of the RBM of each tube is expected in dimer [10]. For long free-standing index-identified SWNTs, prepared by using the same CVD method as the one used to synthesize SWNT dimer, we have established the following RBM frequency vs. diameter relation [5]:

$$\omega_{RBM} (d) = 227/d *(1+Cd^2)^{0.5} \text{ with } C=0.065 \text{ nm}^{-2}.$$

From this relation, the RBM of the (16,12) SWNT is predicted at 129 cm^{-1} and the RBM of the (15,11) SWNT at 142 cm^{-1}. As expected from calculations, both pairs of BM are centered at frequencies close to the RBM frequencies of the (15,11) and (16,12) SWNT respectively. Note that the use of the relation: $\omega_{RBM} (d) = 228/d$, found by Liu and al. [11], lead to an unrealistic upshift (of about 10 cm^{-1}) of the BMs with respect to the RBMs.

By using the spectral moment method [9], we have been able to calculate the non-resonant Raman spectrum of the (15,11)-(16,12) dimer. For this purpose, the distance between the nearest carbon atoms of interacting tubes was fixed at 0.34 nm. The result obtained is shown on figure 1.b. To compare calculations with experimental data, the calculated frequencies were rigidly up-shifted by 13 cm^{-1} (in order to take into account the presence of the additional term in the ω_{RBM} (d) relation). Furthermore, the intensity of the high frequency modes was also adjusted (reduced) in order to reproduce the profile of the experimental spectrum. To justify this procedure, we assume that the high-frequency modes are non-resonant at 1.92 eV, and their presence in the spectrum, with a relative weak intensity, is a consequence of the coupling with the resonant low-frequency modes [8]. This experimental result supported by theoretical calculations highlights that new breathing modes can be present in the Raman spectrum of SWNT dimers.

Raman spectroscopy of index-identified double-walled carbon nanotubes

The Raman spectra of six individual, free-standing, and index-identified DWNTs have been measured (information about their structure is summarized in table 1). These DWNTs have different interlayer distances ranging from 0.315 nm to 0.365 nm.

#	Indices	Diameter (nm)	Type	$\Delta d/2$ (nm)
1	(12,8)@(16,14)	1.37@2.04	SC@SC	0.335
2	(18,2)@(20,12)	1.49@2.19	SC@SC	0.35
3	(13,9)@(24,7)	1.50@2.21	SC@SC	0.355
4	(22,11)@(27,17)	2.28@3.01	SC@SC	0.365
5	(23,5)@(22,17)	2.03@2.65	M@SC	0.315
6	(34,10)@(47,3)	3.13@3.80	M@SC	0.34

Table 1: Chiral indices, diameter, type of inner and outer tubes, and interlayer distance ($\Delta d/2$) of the 6 DWNTs investigated in this work.

To illustrate our results on DWNTs, the low-frequency Raman spectra of the (12,8)@(16,14) and (13,9)@(24,7) DWNTs are displayed in figure 2 (top and bottom respectively). Two well-defined peaks appear in the both spectra. These modes are assigned to the collective in-phase (at low-frequency) and out-of-phase (at high-frequency) breathing modes of both concentric layers, the so-called radial breathing-like modes (RBLMs) [12]. The frequencies of the individual free-standing (12,8) SWNT [5] and (13,10) SWNT [6] are indicated by dashed lines. An appreciably blue-shift of RBLMs is observed in DWNTs with respect to SWNT's RBM. These shifts indicate that the relationships, $\omega_{RBM}(d)$, established for SWNTs are not valid for DWNTs. In other words, the use of the different $\omega_{RBM}(d)$ relationships established for SWNTs to derive the diameters of the inner and outer tubes from the in-phase and out-of-phase RBLM respectively leads to a systematic underestimation of both diameters.

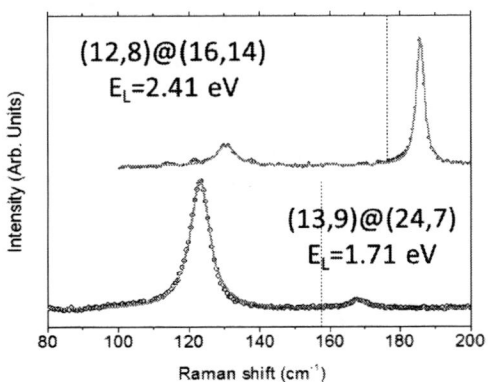

Figure 2: The low-frequency range of the Raman spectra of the (12,8)@(16,14) DWNT (top) and (13,9)@(24,7) DWNT (bottom) excited at 2.41 eV and 1.71 eV respectively. The experimental frequencies of the individual (12,8) SWNT [5] and (13,10) SWNT [6] (close in diameter to the (13,9) SWNT) are indicated by dashed-lines.

Different authors have calculated the dependence of the RBLMs frequencies with the diameter of the tubes [see for example Ref.11 and references therein]. In these calculations, van der Waals interactions between the tubes are introduced and the interlayer distance is fixed close to 0.34 nm.

Figure 3 (solid lines) displays the theoretical dependence of the in-phase (left) and out-of-phase (right) RBLM as a function of the outer tube diameter [12,13]. The experimental frequencies of the in-phase and out-of-phase RBLMs of the DWNTs listed in Table 1are also reported on the same figure (Fig. 3, symbols). Focusing on the out-of-phase RBLM, three behaviors are evidenced: (i) when the interlayer distance is close to 0.34 nm (Fig.3(right), green dots) the experimental frequencies fit the calculated ones, (ii) for interlayer distance larger than 0.34 nm the experimental RBLM are at lower frequencies than the calculated ones (Fig.3 (right), blue squares): more precisely, the larger the interlayer distance, the larger the difference between experiment and calculation. (iii) For the tube with an interlayer distance smaller than 0.34 nm, the RBLM is at higher frequency than the theoretical prediction (Fig.3 (right), red diamond). These results illustrate the strong dependence of the out-of-phase RBLM frequencies on the interlayer distance or in other words on the strength of the inner tube-outer tube interaction. For the in-phase RBLM (Fig.3, left), the experimental frequencies match the calculations without a clear signature of the role of the interlayer distance. The difference between calculations and experiments is smaller than ± 3 cm^{-1}. With regards to these observations, the positions of both RBLMs and the gap between experiment and calculations can be used as criteria to evaluate diameters of the inner and outer tubes in non-index-identified DWNTs. The outer diameter is derived from best matching between the experimental and calculated frequencies of the in-phase RBLM, and the interlayer distance is evaluated from the difference between the experimental and calculated frequencies of the out-of-phase RBLM.

Recently, by combining electron diffraction and Raman spectroscopy on 13 index-identified DWNTs, Liu et al. [11] obtained results in close agreement with those presented here. In addition, from the evaluation of the average unit-area force constant for all their DWNTs and from the comparison of these average force constants with high-pressure graphite measurements, these authors have been able to connect the interlayer distance with the internal effective pressure. They found that for interlayer distance larger (smaller) than 0.34 nm a negative (positive) pressure occurs between the tubes. The extrapolation of their approach to the present measurements leads to estimate a pressure ranging from 2 GPa to -1.75 GPa, for interlayer distance between 0.315 nm and 0.365nm.

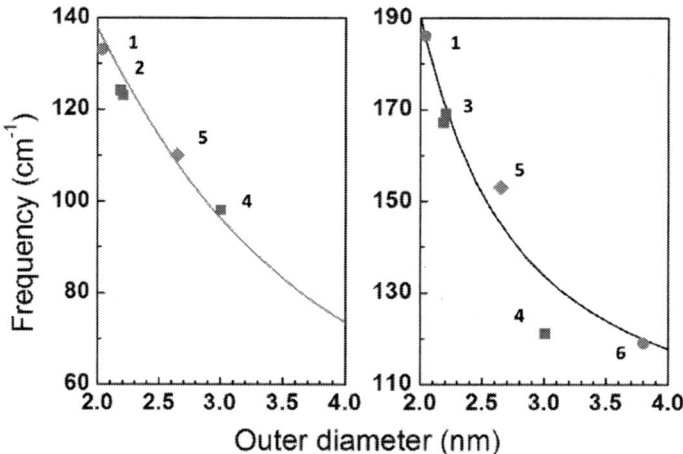

Figure 3: Comparison between calculated and experimental RBLM frequencies. (left: in-phase RBLM and right: out-of-phase RBLM).Calculations, (solid lines); RBLMs of DWNTs having an interlayer distance close of 0.34 nm (green dots); RBLMs of DWNTs having an interlayer distance larger than 0.34 nm (blue squares); RBLMs of a DWNTs having an interlayer distance of 0.315 nm (red diamond). The numbers refer to table 1.

Another important conclusion of this study concerns the resonance conditions of the RBLMs. As shown in figure 2, the in-phase and out-of-phase RBLMs of both DWNTs are observed on the same spectrum and for the same excitation energy. The resonance conditions of each RBLM, and correspondingly the optical transitions of each DWNT, were derived from the measurements of Raman spectra using a large number of incident wavelengths. An example of such measured excitation profile of the in phase and out-of-phase RBLM in the case of the (13,9)@(24,7) DWNT is displayed in figure 4 (left).

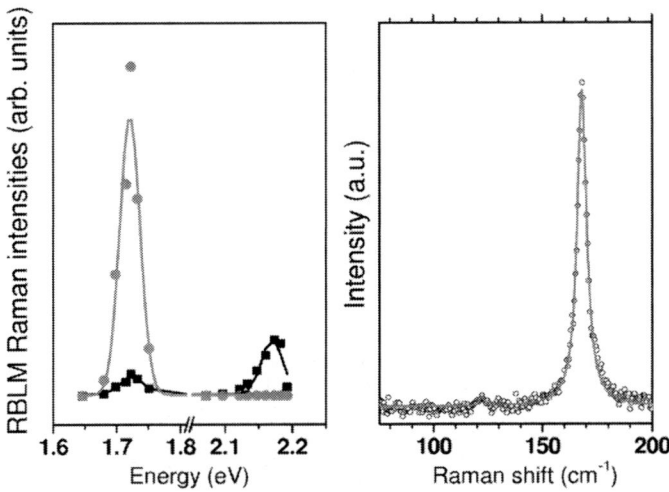

Figure 4: (left) Excitation profile of the in phase (dots) and out-of-phase (squares) RBLM for the (13,9)@(24,7) DWNT. (right) The low-frequency range of the Raman spectra of the (13,9)@(24,7) DWNT excited at 2.17 eV.

For the (12,8)@(16,14) DWNT (Fig. 2), the out-of-phase RBLM shows a maximum intensity when it is excited at 2.41 eV close to the E_{33} transition (2.43 eV) expected for the (12,8) inner tube. By contrast no transition is expected at this energy for the (16,14) outer tube.

For the (13,9)@(24,7) DWNT, the in-phase RBLM shows a maximum intensity when it is excited at 1.71 eV (Fig.2), i.e. about 100 meV lower than the E_{33} transition (1.83 eV) expected for the (24,7) SWNT(no transition is found close to 1.71 eV for the (13,9) SWNT). The out-of-phase RBLM shows a maximum of its intensity when it is excited at 2.17 eV (Fig.4, right), i.e. about 100 meV lower than the E_{33} transition (2.27 eV) expected for the (13,9) SWNT (no transition is expected at 2.17 eV for the (24,7) SWNT).

From this investigation, we conclude that both coupled RBLMs can be resonantly excited even if one electronic transition of either one of the walls matches the excitation energy.

In addition, for an interlayer distance close to 0.34 nm, the results suggest that only a slight shift of the normalized Kataura plot, (established for SWNTs [3,5]), is necessary to match the optical transitions in DWNT (as illustrated from the results obtained on the (12,8)@(16,14) DWNT) [12]. In contrast, for interlayer distances larger than 0.34 nm, a downshift of several tenths of meV of the Kataura plot is necessary to match the experimental and calculated transition energies (as illustrated from the results obtained on the (13,9)@(24,7) DWNT). These results suggest that the amplitude of the optical transition energies shift within a DWNT compared to its constituting individual SWNT could depend on the interlayer distance. A systematic study of the amplitude of this shift on a large number of index-identified DWNTs, having a broad dispersion of interlayer distances, has still to be performed in order to confirm this hypothesis.

CONCLUSIONS

We presented in this paper a Raman study of two van der Waals-coupled carbon nanostructures, namely, a bundle of two non-identical SWNTs (inhomogeneous dimer) and double- walled carbon nanotubes (DWNTs).

The Raman spectrum of an individual index-identified inhomogeneous dimer, the (15,11)-(16,12) dimer, is reported. In agreement with calculations, the spectrum displays four BM peaks: a lower-frequency pair of peaks around the RBM position of the (16,12) SWNT and a higher-frequency pair of peaks around the RBM position of the (15,11) SWNT. The splitting of the modes is due to the lowering of the symmetry in dimers. Each BM peak can be associated to a mixing of RBM and E modes of the isolated SWNTs.

Investigations of index-identified DWNTs permit to address definitively the following points: (i) The relationships, $\omega_{RB}(d)$, established for SWNTs are not valid for DWNTs. (ii) The out-of-phase RBLM frequencies depend on the interlayer distance or in other words on the strength of the tube-tube interaction. In agreement with Liu et al. [11], it is possible to connect interlayer distance to an internal effective pressure between the layers. (iii) Both coupled RBLMs will be resonantly excited if an electronic transition of either wall matches the excitation energy.

This work illustrates the complexity in the mechanisms governing the resonant Raman responses of coupled nano-systems based on carbon nanotubes.

ACKNOWLEDGMENTS

D. Levshov and X. T. Than acknowledge French government for financial support. This work has been done in the framework of the GDR-I "Science and Application of the Nanotubes".

REFERENCES

1. J.C. Meyer, M. Paillet, T. Michel, A. Moreac, A. Neumann, G.S. Duesberg, S. Roth, and J.-L. Sauvajol, Phys. Rev. Lett. **95**, 217401 (2005)
2. M. Paillet, T. Michel, J.C. Meyer, V.N. Popov, L. Henrard, S. Roth, and J.-L. Sauvajol, Phys. Rev. Lett. **96**, 257401 (2006)
3. T. Michel, M. Paillet, J.C. Meyer, V.N. Popov, L. Henrard, and Sauvajol J-L, Phys. Rev. B **75**, 155432 (2007)
4. T. Michel, M. Paillet, J.C. Meyer, V.N. Popov, L. Henrard, P. Poncharal, A.A. Zahab, and J.-L. Sauvajol, Physica Status Solidi b **244**, 3986 (2007)
5. T. Michel, M. Paillet, D. Nakabayashi, M. Picher, V. Jourdain, J.C. Meyer, A.A. Zahab, and J.-L. Sauvajol, Phys. Rev. B **80**, 245416 (2009)
6. R. Pfeiffer, F. Simon, H. Kuzmany, and V. Popov, Phys. Rev. B **72**, 161404 (2005)
7. J.-C. Blancon, M. Paillet, H. N. Tran, X. T. Than, S. Aberra Guebrou, A. Ayari, A. San Miguel, N.-M. Phan, A.-A. Zahab, J.-L. Sauvajol, N. Del Fatti, and F. Vallée, Nat. Comm. **4**, 2542 (2013) and references therein

8. A. Débarre, M. Kobylko, A.-M. Bonnot, A. Richard, V.N. Popov, L. Henrard, and M. Kociak, Phys. Rev. Lett. **101**, 197403 (2008)
9. K. Sbai, A. Rahmani, H. Chadli, and J.-L. Sauvajol, J. Phys.: Condens. Matter **21**, 045302 (2009)
10. L. Henrard, V. Popov, and A. Rubio, Phys. Rev. B **64**, 205403 (2001)
11. K. Liu, W. Wang, M. Wu, F. Xiao, X. Hong,S. Aloni, X. Bai, E. Wang, and F. Wang, Phys. Rev. B **83**, 131404 (2011)
12. D. Levshov, T. X. Than, R. Arenal, V. N. Popov, R. Parret, M. Paillet, V. Jourdain, A. A. Zahab, T. Michel, Yu. I. Yuzyuk, and J.-L. Sauvajol, Nano Lett., 11 (11), 4800 **(2011)**
13. V.N. Popov and L. Henrard, Phys. Rev. B **65**, 235415 (2002).

Mater. Res. Soc. Symp. Proc. Vol. 1700 © 2014 Materials Research Society
DOI: 10.1557/opl.2014.575

Optical properties of nanostructured carbon and gold nanoparticle hybrids

Yuan Li,[1] Nitin Chopra,[1,2,*]

[1]Metallurgical and Materials Engineering Department, Center for Materials for Information Technology (MINT), The University of Alabama, Tuscaloosa, AL 35487, U.S.A.
[2]Department of Biological Sciences, The University of Alabama, Tuscaloosa, AL 35487, U.S.A.
*Corresponding Author E mail: nchopra@eng.ua.edu, Tel: 205-348-4153, Fax: 205-348-2164

ABSTRACT

We report simulation of optical properties of hybrid geometry comprised of multilayer graphene shell encapsulated gold nanoparticles loaded with carbon nanotubes. The discrete dipole approximation (DDA) method was employed. The results indicated that the optical properties of encapsulated gold nanoparticles were not suppressed by the carbon material coating. Furthermore, low scattering effects were also observed. The simulation method helped visualize the near-surface normalized electric field, which is directly related to the intensity of hot spots on the surface of these hybrid nanoarchitectures.

INTRODUCTION

Self-assembly and chemical patterning of carbon nanotubes (CNTs) is critical for applications in nanoelectronics and sensors [1-4]. Recently, the reproducibility and sensitivity of devices based on CNTs for DNA recognition was demonstrated [5]. Complex nanoscale heterostructures based on CNTs are of particular interest to achieve greater functionality in nanodevices and sensors [6,7]. In this direction, controlled arrangement or positioning of CNTs on complex substrates is still a big challenge. Recently, the authors demonstrated a facile and scalable chemical vapor deposition (CVD) approach for the growth of multilayer graphene shells on gold (Au) nanoparticles (GNPs) [8]. The surface chemistry of multilayer graphene shell was demonstrated for applications in bioanalysis. Here, we report the basic simulation results of hetero-nano-architectures comprised of CNTs and GNPs (GNP-CNT heterostructures). The simulation was performed using discrete dipole approximation (DDA) approach.

EXPERIMENT

Discrete dipole approximation (DDA) method was used to simulate the optical properties of GNP-CNT heterostructures. This method is based on 3-D Maxwell equation via the DDSCAT code developed by Draine and Flatau [9,10]. Detailed simulation procedures have been demonstrated by our group in an earlier report [11]. Briefly, DDSCAT 7.2 was utilized to estimate normalized electric field intensity or surface plasmon generation for the GNP-CNT heterostructures. The normalized electric field intensity is defined as the ratio between the electric field generated near the nanostructures and the incident electric field ($|E|/|E_0|$). The effective radius of dipoles was calculated using: $R_{eff} = (3V/4\pi)^{1/3}$, where V is total volume of the material in the target and is calculated using $V = Nd^3$ (N is the number of dipoles and d is the lattice spacing in cubic array).

DISCUSSION

Figure 1 shows the HRTEM image of GNPs fabricated in CVD process developed by the authors [8]. These as-produced GNPs have an average size of ~72.19 ± 10.17 nm. The thickness of multilayer graphene shells was ~3.8 nm and the c-axis lattice spacing was observed to be ~0.38 nm. This value is slightly higher than that for c-axis spacing of graphite (~0.34 nm) due to the curvature in the multilayer graphene shells [10].

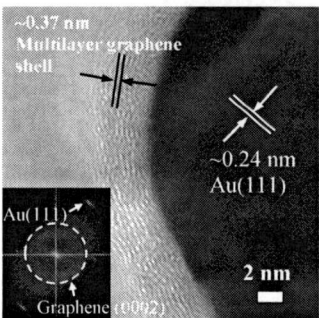

Figure 1. HRTEM image and corresponding FFT of the as-produced GNPs.

To understand the optical properties of GNP-CNT heterostructures, DDSCAT was used to simulate their surface extinction and normalized electric field distribution. The nanostructured configurations (or target) in this study were comprised of an arrangement of dipoles for which extinction spectra and electric field distributions were numerically solved. The extinction efficiency is the combined effect of absorbance and scattering [44]. In the simulation, the target was assumed to be composed of one GNP linked with 25 CNTs (Figure 2A-C). The Au nanoparticles core was assumed to be 50 nm (diameter) and the encapsulated multilayer graphene shell was assumed to be 0.8 nm thick. CNT with lengths of ~10 nm, outer diameter of 10 nm and inner diameter of ~5 nm were considered. DDA simulation method helped calculate the extinction spectra and normalized electric field distribution for GNP-SWCNT heterostructure. The obtained extinction efficiency spectra showed the combined effect of scattering and absorption of the incident electromagnetic wave. Figure 2D indicates well-defined extinction peak at ~524 nm [12,13] corresponding to the encapsulated Au nanoparticle within the multilayer graphene shells. This extinction spectrum was further split into absorbance and scattering components. As shown in Figure 2D, absorbance had a dominant contribution to the extinction spectra while the effect of scattering was relatively small. This is because the target size is much smaller than the wavelength of incident wave and scattering falls in Rayleigh regime [14,15]. The scattering (Q_{sca}) can be written as $Q_{sca} = \frac{8\pi}{3} k^4 R^6 F(m)$, where $k = \frac{2\pi}{\lambda}$ (λ is the wavelength), R is the radius of the particles, and $F(m)$ is a function of the refractive index (m). The absorption (Q_{abs}) can be written as $Q_{abs} = 4\pi k R^3 E(m)$, where $E(m)$ is also the function of the refractive index relying on a non-zero imaginary part. Thus, the ratio of scattering to absorbance is $\frac{2}{3}(kR)^3\frac{E}{E}$, which is proportional to $(kR)^3$. Considering the average diameter of the heterostructures as ~72 nm ($R = 36$ nm) and at a wavelength of ~500 nm, the ratio $(kR)^3 = 0.1$. This result shows a good agreement with the ratio of scattering to absorbance in figure 2D.

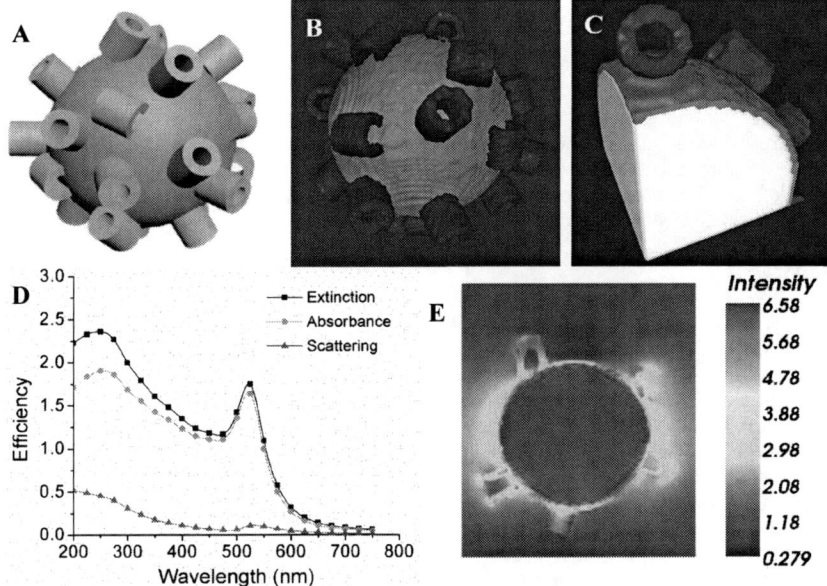

Figure 2. (A) 3-D image of the GNP-CNT heterostructures. (B, C) Corresponding 3-D images showing sections of GNP-CNT heterostructures. (D) Calculated extinction, absorbance, and scattering plots as a function of wavelength for a GNP-CNT heterostructure target. (E) Normalized electric field distribution on a GNP-CNT heterostructure target at the incident wavelength of ~530 nm.

On further comparison with bare Au nanoparticles [11], this low scattering was also attributed to the presence of multilayer graphene shell as well as CNTs. Moreover, the nanostructures were almost single crystalline and thus, very few scattering sites existed in the heterostructures. The interfaces could also be the major sources of scattering phenomena. In addition, it is noteworthy that another broad extinction peak was observed in the UV region. This peak has been assigned to the interband transition of d-band electrons from Au nanoparticles [12,16], which is mainly due to the excitation of d electrons when the energy of incident photon is larger than the gap energy. This direct interband extinction can occur at any energy-conserved position in the brillouin zone. The energy conservation equation is given by:

$$\hbar\omega = \varepsilon_c(k) + \varepsilon_v(k) \qquad (1)$$

where $\hbar$ is the Planck's constant, ω is the frequency. $\varepsilon_c(k)$ and $\varepsilon_v(k)$ are the dielectric function of the empty band and full band. Additionally, it has been observed that the low wavelength (UV region) peak is intensified due to the presence of multilayer graphene and could be attributed to multilayer graphene shell assisting transition of Au d-band electrons. The peak wavelength of the extinction spectrum was further selected as incident wavelength to generate the normalized electric field distribution. As indicated by Figure 2E, the intense electric field (generally called "hot spots") was mainly observed at the contact interface of CNTs and GNP. This further confirmed the efficient interaction of surface plasmon of Au nanoparticle with surrounding carbon nanostructures.

CONCLUSIONS

The optical properties of GNP-QD heterostructures were simulated for understanding absorbance and scattering phenomena as well as surface electric field distribution using DDA method. The result indicated significantly suppressed scattering for GNP-CNT heterostructures and intense normalized electric field at the interface of CNTs and GNP.

REFERENCES

1. K. Keren, R. S. Berman, E. Buchstab, U. Sivan, and E. Braun, *Science* **302**, 1380 (2003).
2. K. A.Williams, P. T. Veenhuizen, G. Beatriz, and C. Dekker, *Nature* **420,** 761 (2002).
3. L. Goux-Capes, A. Filoramo, and J. N. Patillon, *AIP Conf. Proc.* **725,** 17 (2004).
4. H. W. C. Postma, T. Teepen, Z. Yao, and C. Dekker, *Science* **293**, 76 (2001).
5. A. Bachtold, P. Hadley, T. Nakanishi, and C. Dekker, *Science* **294**, 1317 (2001).
6. D. J. Hornbaker, S. J. Kahng, S. Misra, and A. Yazdani, *Science* **295**, 828 (2002).
7. Z. Yao, N. Braidy, G. A. Botton, and A. Adronov, *J. Am. Chem. Soc.* **125**, 16015 (2003).
8. N. Chopra, L. G. Bachas, and M. Knecht, *Chem. Mater.* **21** 1176 (2009).
9. P. J. Flatau, and B. T. Draine, *J. Opt. Soc. Am. A* **11**, 1491 (1994).
10. B. T. Draine, and P. J. Flatau, *arXiv preprint arXiv* **1305**, 6497 (2013).
11. J. Wu, W. Shi, and N. Chopra, *Carbon* **68**, 708 (2014).
12. J. A. Creighton, and D. G. Eadon, *J. Chem. Soc., Faraday Trans.* **87**, 3881 (1991).
13. S. Zhu, T. P. Chen, Y. C. Liu, and S. Fung, *J. Nanopart. Res.* **14**, 1 (2012).
14. P. K. Jain, K. S. Lee, I. H. El-Sayed, and M. A. El-Sayed, *J. Phys. Chem. B* 110, 7238 (2006).
15 C. F. Bohren, and D. R. Huffman, *John Wiley & Sons*. 130, (2008)
16. K. Yamada, K. Miyajima, and F. Mafuné, *J. Phys. Chem. C,* **111**, 11246 (2007).

Chemical and Biological Investigation

Mater. Res. Soc. Symp. Proc. Vol. 1700 © 2014 Materials Research Society
DOI: 10.1557/opl.2014.850

A Continuous Flow Device for the Purification of Semiconducting Nanoparticles by AC Dielectrophoresis

Rustin Golnabi[1], Su (Ike) Chih Chi[1], Stephen L. Farias[1], and Robert C. Cammarata[1, 2]

[1]Department of Materials Science and Engineering, Johns Hopkins University, 3400 N. Charles Street, Baltimore, Maryland 21218, USA

[2]Department of Mechanical Engineering, Johns Hopkins University, 3400 N. Charles Street, Baltimore, Maryland 21218, USA

ABSTRACT

Single-walled carbon nanotubes (SWCNTs) have attracted significant attention as building blocks for future nanoscale electronics due to their small size and unique electronic properties. However, current SWCNT production techniques generate a mixture of two types of nanotubes with divergent electrical behaviors due to structural variations. Some of the nanotubes act as metallic materials while others display semiconducting properties. This random mixture has prevented the realization of functional carbon nanotube-based nanoelectronics. Here, a method of purifying a continuous flow of semiconducting nanotubes from an initially random mixture of both metallic and semiconducting SWCNTs in suspension is presented. This purification uses A/C dielectrophoresis (DEP), and takes advantage of the large difference of the relative dielectric constants between metallic and semiconducting SWCNTs. Because of a difference in magnitude and opposite directions of a dielectrophoretic force imposed on the random SWCNT solution, metallic SWCNTs deposit onto an electrode while semiconducting SWCNTs remain in suspension [3]. A discussion of these techniques is presented, along with a dielectrophoretic force-utilized microfluidic lab-on-a-chip device that can accomplish purification of semiconducting nanoparticles at high processing rates. The effectiveness of the device is characterized using Raman spectroscopy analysis on separated samples.

INTRODUCTION

Nanoparticles have continued to be at the forefront of electronics research due to their unique properties and size [1-4]. Semiconducting nanoparticles are especially useful for their various applications in optoelectronics, transistors, biosensors, and drug delivery [5-7]. One of the most investigated and promising nanomaterial is carbon nanotubes. Single-walled carbon nanotubes (SWCNTs) are nanostructures which can be thought of as a single atomic sheet of graphene rolled into a seamless cylinder. They are highly regarded for their high thermal conductivity (over 6000 W/mK), high Young's modulus (~1 TPa), light weight, and unique electronic properties [8-10].

When synthesized, SWCNTs form in a distribution of structures and sizes, defined by the chiral lattice vector (m, n). Depending on their chirality, carbon nanotubes display either metallic or semiconducting properties and typically grow in a 1:3 ratio of metallic to semiconducting [2]. One of the challenges to implementing carbon nanotubes for commercial applications is this fact that these two types are synthesized together [3]. Many applications require high purity semiconducting nanotubes. For example, functional transistor arrays require >99.99% purity [11]. To alleviate this issue, many methods have been investigated to efficiently separate

SWCNTs by electronic type, including ultracentrifugation, chemical methods, and chromatography methods [12-14]. These methods rely on very small differences in size, density, and surface structure of the tubes making cost-effective separation challenging. Techniques developed thus far are often expensive, sometimes limited in scale, and utilize methods that may cause damage to the nanotubes. Alternative to separation systems, significant research has also been conducted to improve selective growth of particular chiralities of nanotubes. While some progress has recently been made towards this goal [15], these methods are highly specific to a given set of chiral structures and have yet to be implemented at commercial levels.

In this paper we present a method of using dielectrophoresis to purify semiconducting carbon nanotubes in large quantities, which has previously been shown to effectively produce small quantities of purified metallic nanotubes from a mixed suspension [12]. Dielectrophoresis relies on the dielectric properties of the nanotubes, and takes advantage of the large difference in intrinsic electronic properties of metallic and semiconducting tubes. We first demonstrate the ability to purify semiconducting single-walled carbon nanotubes (s-SWCNTs) from an initial droplet of mixed SWCNT suspension in a static system. Then, we present a microfluidic lab-on-a-chip device which outputs a continuous stream of purified s-SWCNTs.

EXPERIMENTAL DETAILS

Static System

A surfactant, sodium dodecyl sulfate (SDS) (1 wt.%), was added to a mixed suspension of SWCNTs (Sigma-Aldrich, carbon > 90%, ≥ 70% carbon as SWCNTs) in water, and this was then ultra-sonicated for 30 minutes and centrifuged at 3000rpm for 2 hours. (It is noted that this preparation may allow for bundles of tubes to remain in suspension and that more aggressive mixing, sonication, and centrifugation would result in a higher fraction of individual tubes in suspension which could lead to more facile separation.) The supernatant was then collected and used in this experiment. Using electron-beam lithography, two concentric gold electrodes were patterned onto a glass substrate (See Figure 1). 1 mL of the suspension was added to the electrode surface, and a sinusoidal alternating current signal with a frequency of 10 MHz was applied using a signal generator. The suspension was allowed to evaporate under ambient conditions while the applied electrical signal continued. Various positions along the electrode were then analyzed using Raman spectroscopy.

Continuous Flow System

Using UV photolithography, a PDMS microfluidic channel was created over a gold electrode on glass similar to that of the static design. The suspensions of SWCNTs and SDS were created as described for above for the Static System except that they were centrifuged for 24 hours. The suspension was also ultra-sonicated for 30 minutes just prior to running the experiment in order to ensure proper dispersion by the surfactant. The suspension was fed through the channel using a constant flow rate syringe pump at 0.4 mL/h. An applied signal frequency of 500 MHz was used for the duration of the experiment. About 1 mL of processed suspension was collected and analyzed using Raman spectroscopy both the control (mixed, centrifuged CNTs) and the collected sample.

Dielectrophoresis for Particle Separation

Separation of nanograms of metallic SWCNTs from mixed suspension by A/C dielectrophoresis has been shown in previous work [12]. DEP enables us to use the contrasting complex dielectric constants of the metallic and semiconducting materials an electric field gradient to propel some nanoparticles in one direction, and others in the opposite direction. The A/C electric field, E, will induce a dipole moment in the nanoparticles. The polarized particles will move in suspension due to a dielectrophoretic force, F_{DEP}, proportional to the real part of the Claussius-Mossotti factor, K, and the gradient of the electric field squared.

$$F_{DEP} \propto Re(K)\nabla|E|^2 \tag{1}$$

K is dependent on the complex dielectric permittivities of the particle and medium, ϵ_p^*, and ϵ_m^*, respectively, as well as L, a shape factor relevant in the context of nanotubes and like-shaped nanoparticles.

$$K = \frac{\epsilon_p^* - \epsilon_m^*}{\epsilon_m^* + L(\epsilon_p^* - \epsilon_m^*)} \tag{2}$$

These complex permittivities are functions of the real permittivity, the conductivity, and the frequency of the applied field.

$$\epsilon^* = \epsilon - \frac{i\sigma}{\omega} \tag{3}$$

At a given frequency, the complex permittivity of a particle may be greater than or less than that of the medium. Because of the difference in complex permittivities between semiconducting and metallic nanotubes it is possible at sufficiently high frequencies to have significant differences in magnitude and direction of the DEP force experienced by different tubes. This enables separation by electronic type.

DISCUSSION

Static Separation

To determine the purity and electronic properties of the SWCNTs deposited onto the electrode, Raman spectroscopy was performed. The spectra are shown in Figure 1, where the green line is from measurements on the inner electrode, and the violet line is from measurements on the outer electrode. There are several relevant regions of interest in the Raman Spectra of SWCNTs. The G-band (~ 1600 cm^{-1}) corresponds to the tangential vibration of the carbon atoms, and the D-band (~ 1300 cm^{-1}) is a typical indicator of amorphous carbon structures. When paired, the ratio of the intensities of these bands is commonly used to show impurities in a given sample of SWCNTs. When comparing the spectra of the two electrodes, we found that the ratio of the D-band intensity to G-band intensity was significantly lower on the outer electrode. This indicates high concentrations of pristine SWCNTs and negligible amounts of other carbonaceous material.

The mixed suspension of carbon nanotubes was centrifuged for 2 hours meaning much unwanted carbon content and catalyst was still present in the suspension before any separation. It was determined that at low frequencies (10 MHz), purification of nanotubes occurs, where SWCNTs are refined from other carbon content that are byproducts of the manufacturing process.

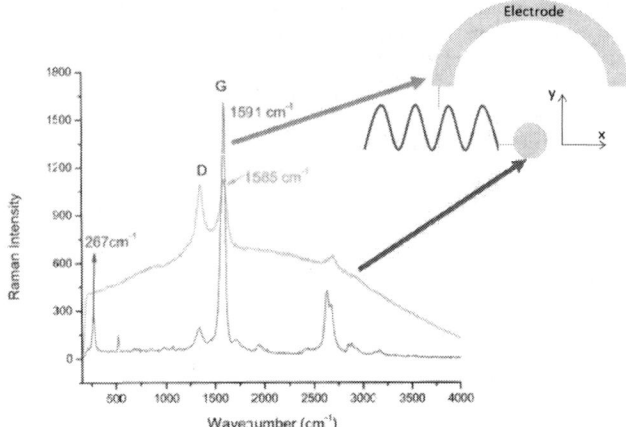

Figure 1. Raman spectra of inner (green) and outer (violet) electrodes of static system operated at 10 MHz: The D and G bands show that this DEP process refines SWCNTs from other carbon content.

Continuous Flow Separation

The microfluidic device was designed to ensure that the strength of the electric field was great enough to propel the metallic nanotubes to attach to the smaller circular electrode. The rest of the nanotubes flow through the device and the output would hence have a relatively high concentration of semiconducting SWCNTs. Before running the experiment, the suspension was centrifuged for 24 hours at 3000 rpm, in order to ensure the purity of SWCNTs and reduce amorphous carbon material. Preliminary results for the system operated at a frequency of 500 MHz, and a fluid flow rate of 0.4 mL/hour are described below and were analyzed using Raman spectroscopy, followed by UV-vis spectroscopy.

The Raman spectra (Figure 2) shows the Radial Breathing Mode (RBM) of the prepared control suspension and the processed sample. The RBM frequency is dependent on tube diameter and can be correlated to a particular chirality and thus the electronic property of the tube. The peak present at around ~260 cm^{-1} corresponds to an (8, 5) chirality, which is metallic. The peak at around ~270 cm^{-1} corresponds to a (10, 2) chirality, which is semiconducting. The shift in RBM profile between the control and processed sample is indicative of semiconducting enrichment.

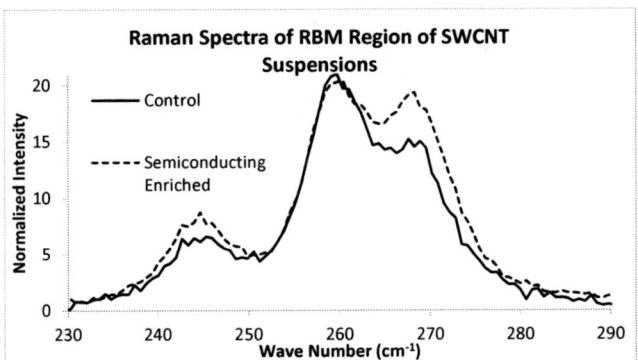

Figure 2. Raman spectra of the radial breathing modes (RBM) of the SWCNT outputs and control. The peak at 260 corresponds to metallic (8,5) while the peak at 270 is semiconducting (10,2). The relative strength (area) of the peaks indicates semiconducting enrichment compared to the control sample.

CONCLUSIONS

Purification of electronic types of SWCNTs by dielectrophoresis was shown to be effective and promising as a bulk processing technique. The microfluidic device developed showed preliminary purification of semiconducting material. In order to begin increasing this purity to needed levels for applications such as biosensors and transistors, our future work includes optimization of the various factors, including flow rate, frequency, and channel design.

ACKNOWLEDGMENTS

We would like to acknowledge funding support for this work from the Maryland Innovation Initiative grant. We would also like to acknowledge the intellectual and experimental support of Professor Chia-Ling Chien, Danru Qu, and Dr. Natalia Drichiko of the Johns Hopkins University Department of Physics and Astronomy.

REFERENCES

1. P. Avouris, *Physics World* **20**, 40-45 (2007).
2. P. Avouris, and J. Appenzeller, *The Industrial Physicist*, June/July 2004, American Institute of Physics.
3. R. Krupke, et al., *Science* **301**, 344-347 (2003).
4. N. Peng, et al., *J. Appl. Phys.* **100**, 024309 (2006).
5. D.V. Talapin, J.S. Lee, M.V. Kovalenko, and E.V. Shevchenko, *Chem. Rev.* **110 (1)**, 389-459 (2010)
6. B. L. Allen, P. D. Kichambare, and A. Star, *Advanced Materials* **19 (11)**, 1439-1451 (2007).
7. M. Liong, et al., *ACS Nano* **2 (5)**, 889-896 (2008).
8. S. Berber, Y-K. Kwon, and D. Tománek, *Phys. Rev. Lett.* **84**, 4613 (2000).
9. N. Yao, and V. Lordi, *J. Appl. Phys.* **84**, 1939 (1998).

10. S.H. Jin, *Nature Nanotechnology* **8**, 347-355 (2013).
11. M.S. Arnold, et al., *Nature Nanotechnology* **1**, 60-65 (2006).
12. R. Krupke, et al., *Nano Lett.* **3**, 1019-1023 (2003).
13. H. Liu, et al., *Nature Comm.* **2**, (2011).
14. R. Krupke, et al., *Nano Lett.* **4**, 1395-1399 (2004).
15. F. Yang, et al., Nature **510**, 522-524 (2014).

Mater. Res. Soc. Symp. Proc. Vol. 1700 © 2014 Materials Research Society
DOI: 10.1557/opl.2014.553

Quadruple Hydrogen Bonded Nanocarbon Networks for High Performance Dispersant-Free Conducting Pastes

Joong Tark Han*, Jeong In Jang, Sua Choi, Seon Hee Seo, Seung Yol Jeong, Hee Jin Jeong, Geon-Woong Lee
Nano Carbon Materials Research Group, Creative and Fundamental Research Division, Korea Electrotechnology Research Institute, 12, Bulmosan-ro 10beon-gil, Seongsan-gu, Changwon 642-120, South Korea

ABSTRACT

Colloidal dispersion of nanocarbon (NC) materials in dilute solutions or pastes is prerequisite for applications of NC-based electrodes from flexible electronics and flexible conducting fibers to electrochemical devices. Here, we show a straightforward method for fabricating NC suspensions with >10% weight concentrations in absence of organic dispersants. The method involves introducing supramolecular quadruple hydrogen bonding motifs into the NC materials without sacrificing the electrical conductivity.

INTRODUCTION

Unfortunately, unless the nanocarbon (NC) materials are highly functionalized with oxidative moieties or are shortened using ball milling, they agglomerate irreversibly at weight concentrations higher than 1.0 wt% because of their high aspect ratio (>500) and strong van der Waals attraction, making further processing difficult [1]. However, such severe covalent functionalization of NC materials inevitably degrades their electrical and/or electronic properties to some degree. Thus, many additives such as organic surfactants, alkali solutions, superacids, N-methylpyrrolidone, and ionic liquids have been used to prepare highly concentrated NC suspensions. A disadvantage of this approach is that the overall properties (e.g., the electrical and thermal properties) of the mixtures can suffer because of high intertube or sheet-to-sheet junction resistance, because the conducting nanomaterials may be separated by the insulating organic materials added to facilitate dispersion and processing. Recently, we have reported several strategic methods for dispersion of chemically modified graphene (CMG) nanosheets in dilute solutions by using titania precursor [2], cation-pi interaction [3] and silanol molecules [4] as shown in Figure 1. We could minimize the conductivity decrease of CMG films after film formation. However, our previous method also does not promise the dispersion of NC materials in highly concentrated solutions, so-called pastes.

Here, we present that quadruple hydrogen bond (QHB) networks can overcome these issues for fabricating printable, spinnable, and chemically compatible conducting pastes containing high quality CNTs and graphene nanoplatelets in organic solvents without the need for additional dispersion agents. The dispersant-free pastes could form rationally designed hybridized materials including polymers, metals, and metal oxides.

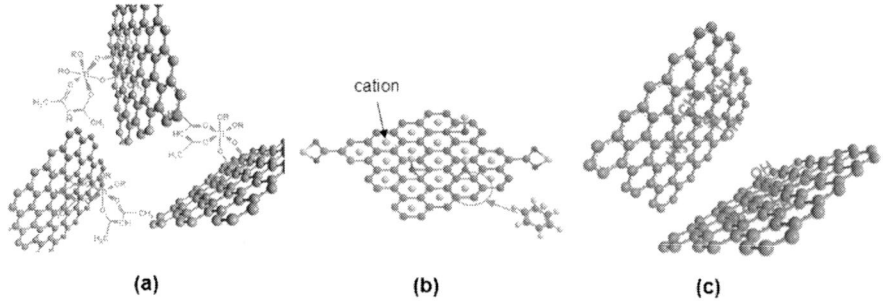

(a)   **(b)**   **(c)**

Figure 1. Schematic illustrations of colloidal dispersion of chemical modified graphene nanosheets in dilute solution assisted by (a) titania precursor, (b) cation-pi interaction, and (c) trimethylsilanol

NC-COOH   **NC-NCO**   **supra-NC**

Figure 2. Synthetic scheme for supra-NCs, in which NCs functionalized with carboxylic acid groups (NC-COOH) were sequentially reacted with toluene diisocyanate (TDI) and 2-amino-4-hydroxy-6-methyl-pyrimidine (AHMP) to form 2-ureido-4[*1H*]pyrimidinone moieties.

RESULTS AND DISCUSSION

QHB arrays have been predicted to be much stronger than triple hydrogen bond arrays [5]. Motivated by a self-assembling DDAA array of QHB donor (D) and acceptor (A) sites [6], we functionalized the NC materials with 2-ureido-4[1H]pyrimidinone moieties, having QHBs as shown in Figure 2. The supra-NCs were synthesized in three steps from commercially available starting materials, carboxylated NCs (CNTs-COOH or graphene-COOH), toluene diisocyante (TDI), and 2-amino-4-hydroxy-6-methyl-pyrimidine (AHMP) [7]. Pristine multi-walled carbon nanotubes (MWNTs) and single-walled carbon nanotubes (SWNTs) were functionalized with carboxylic acid using mixture of sulfuric acid and nitric acid (7/3, volumetric ratio) at low temperature, 50 °C to minimize the defect formation for 24 h, and dispersed in dimethylforamide (DMF) by bath sonication for 1h. Graphene oxide (GO) nano-platelets having carboxylic acid groups were prepared by oxidation of graphite with modified Hummers method followed by exfoliation in DMF by bath sonication for 1 h.

We could not prepare the NC-based paste with NCs just functionalized with carboxylic acid in organic solvents by severe agglomeration without dispersant as shown in Figure 3, which makes it less suitable for further applications. However, using our approach of supramomolecular structure based on QHB motifs, we show the achievement of well-dispersed pastes that have fluid surface tensions low enough to spread on a wide range of substrates yet high enough to prevent colloidal agglomeration after deposition. It is particularly worth noting that supra-NCs interconnected with QHB motifs were formed the well-dispersed pastes even after a simple stirring with 10% concentration. Supra-MWNTs show the interconnected micro-ball structures as observed in dried dilution solution, which is abnormal structures of functionalized MWNTs. Interestingly, this image is very similar to asian fermented soybeans.

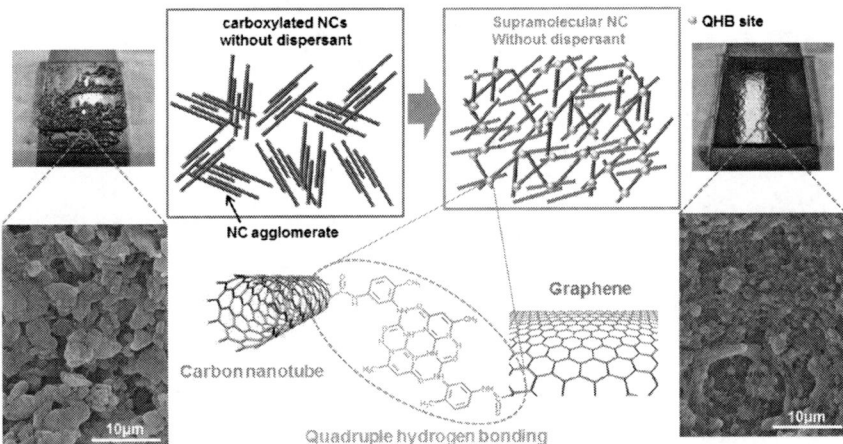

Figure 3. Photo images of unstable colloidal suspension of carboxylated MWNTs (left) and stable supra-MWNT pastes without dispersant (right). The schematic illustration shows the dispersion mechanism by quadruple hydrogen bonding (center). The interconnected microball structure in the left lower SEM image shows the QHB networking of supra-MWNTs.

Using these stable supra-NC pastes, we investigate the rheological properties to demonstrate the strong H-bonding effect in paste by measuring the shear viscosity. In hard-sphere colloidal suspensions, when the volume fraction of a colloidal suspension is increased upon beyond ~50%, its low-shear viscosity increases often displays a shear-thickening or jamming transition. Dilation within a fixed volume of suspending liquid has been suggested as a possible mechanism for a shear-thickening in concentrated colloidal suspensions. The details of this behaviour can be also dependent on the nature of the interactions between particles.[8] If the particles are interconnected or assembled with multivalent interaction motifs, even in low concentration liquids below 10%, we presumed that the viscosity of the liquid can be increased at low shear region. Figure 4a shows the shear viscosities of supra-MWNT, supra-SWNT and supra-graphene pastes stabilized with QHBs as a function of shear rate. Importantly, we could observe the shear thickening phenomena at low shear rate region and then viscosity decreased linearly at high shear rate. These results clearly illustrate the NC materials are interconnected by strong interactions attached on the surface or at the end of CNTs and graphene nanosheets.

Moreover, these unique pastes are printable, spinnable, and could be formed into flexible buckypaper by casting or filtration (Figure 4b-d). Moreover, it is well known that from MWNTs, it is difficult to fabricate the buckypaper using a conventional filtration process without severe oxidation and shortening of the nanotubes because of their large diameter which reduces the strong π-π interactions. These results lay the foundation for building a multifunctional NC-based hybrid materials.

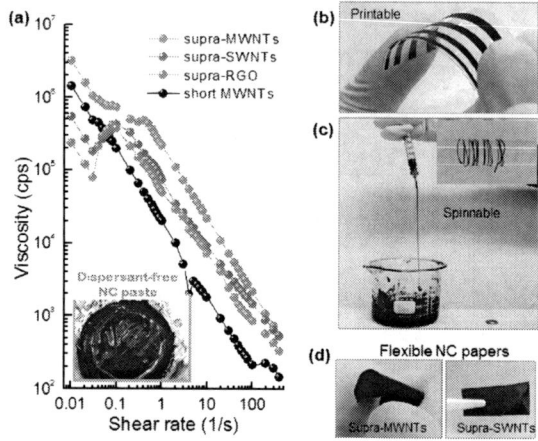

Figure 4. (a) Shear viscosity plot for supra-NC pastes (~10 wt%) in DMF as a function of shear rate showing shear thickening behavior at a low shear rate. In contrast, short MWNTs just showed the typical shear thinning behavior. (b) The printed film on the plastic substrate, (c) The conductive fiber with a high supra-SWNT content (~ 70 wt%), (d) Bucky papers with supra-MWNTs (left) and supra-SWNTs (right).

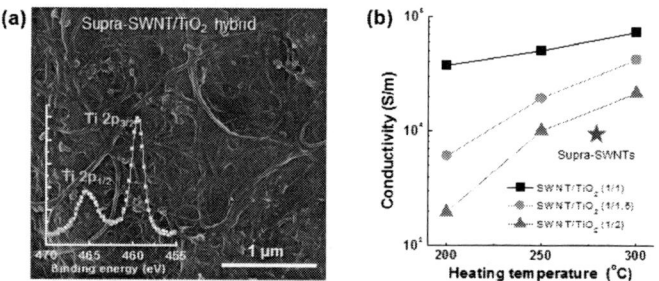

Figure 5. (a) SEM image of supra-SWNT/TiO$_2$ hybrid film. Inset shows XPS spectrum of it indicating existence of TiO$_2$. (b) Conductivity of supra-SWNT/TiO$_2$ hybrid film by varying amount of TiO$_2$ sol as a function of heat treatment temperature.

After that, to investigate the chemical compatibility of the supra-NCs, we hydridized supra-NCs with polymer, metal or metal oxide materials by a simple stirring and in-situ synthesis in solution. First, to improve the mechanical and electrical properties of SWNT coating, polymerized TiO_2 precursor sol (titanium isopropoxide: acetylacetone mixture) in DMF was hybridized with supra-SWNTs. Acetylacetone used as a stabilizer of TiO_2 precursor sol can induce the hydrophobic interaction between SWNT surfaces and TiO_2 sol, resulting in formation of uniform supra-SWNT/TiO_2 hybrid films (Figure 5a). The conductivities of the prepared films were dependent on the amount of TiO_2 sol. The conductivity of supra-SWNT film was dramatically enhanced by adding TiO_2 sol and heat treatment at 300 °C from ~8000 to ~80000 S/m (Figure 3b), which is due to p-type doping effect.

CONCLUSIONS

In summary, we demonstrated that dispersant-free NC paste can be prepared by introducing QHB motifs onto NC materials such as MWNTs, SWNTs, graphene nano-platelets. QHB motifs contribute to form the printable, spinnable, and compatible NC-based conducting pastes. The shear thickening phenomena of supra-NC pastes and self-assembling behavior of supra-NC particles clearly shows the interconnected strutures of supra-NCs by QHB motifs.

ACKNOWLEDGMENTS

This research was supported by the Center for Advanced Soft Electronics under the Global Frontier Research Program of the Ministry of Science, ICT, and Future Planning as Global Frontier Project (CASE-2013M3A6A5073177), and by the Primary Research Program of Korea Electrotechnology Research Institute, South Korea.

REFERENCES

1. A. Thess et al. *Science* **273**, 483–487 (1996).
2. J. T. Han, B. J. Kim, B. G. Kim, J. S. Kim, B. H. Jeong, S. Y. Jeong, H. J. Jeong, J. H. Cho, G. -W. Lee, *ACS Nano* **5**, 8884-8891 (2011).
3. S. Y. Jeong, S. H. Kim, J. T. Han, H. J. Jeong, G. -W. Lee, *Adv. Func. Mater.* **22**, 3307-3314 (2012).
4. J. T. Han, J. I. Jang, B. H. Jeong, B. J. Kim, S. Y. Jeong, H. J. Jeong, J. H. Cho, G. –W. Lee, *J. Mater. Chem.* **22**, 20477-20481 (2012).
5. R. P. Sigbesma et al. *Science* **278**, 1607-1604 (1997).
6. J. T. Han, D. H. Lee, C. Y. Ryu, K. Cho, Fabrication of superhydrophobic surface from a supramolecular organosilane with quadruple hydrogen bonding. *J. Am. Chem. Soc.* **126**, 4796-4797 (2004).
7. J. T. Han et al. *Nature Commun.* **4**, 2491 (2013)
8. B. J. Maranzano, N. J. Wagner, *J. Rheol.* **45**, 1025 (2001).

Mater. Res. Soc. Symp. Proc. Vol. 1700 © 2014 Materials Research Society
DOI: 10.1557/opl.2014.535

Nanotubes and nanoparticles based 3D scaffolds for the construction of high performance Biosensors

Meenakshi Singh[1], Michael Holzinger[1], Maryam Tabrizian[2], and Serge Cosnier[1]

[1]Université Joseph Fourier, Département de Chimie Moléculaire UMR-5250, ICMG FR-2607, CNRS, Grenoble, France.
[2] McGill University, Biomat'X Research Laboratories, Dept. of Biomedical Engineering and Faculty of Dentistry, Montréal, Canada.

ABSTRACT

3D scaffolds with different pore sizes, using single-walled carbon nanotubes (SWCNTs) and nanoparticles of different size were constructed. Biotinylated glucose oxidase (GOX-B) and anti-cholera toxin (anti-CT) were immobilized onto the one and two level nanoscaffolds, functionalized with pyrene-β-cyclodextrin for the construction of glucose based enzyme sensors and immunosensors, respectively. For enzyme sensors, highest current density and sensitivity ($41.72 \ \mu A \ cm^{-2}$, $3 \ mA \ M^{-1} \ cm^{-2}$) were obtained with two level scaffolds made with 100 nm nanoparticles. In contrast to this, for immunosensors, highest current density and sensitivity ($11.71 \ \mu A \ cm^{-2}$, $116.2 \ \mu A \ M^{-1} \ cm^{-2}$) were obtained with two level scaffolds made with 500 nm nanoparticles, indicating that the pore sizes can be adjusted using different size of nanoparticles for the respective applications.

INTRODUCTION

In the past years, several approaches have been reported to realize electrochemical detection of biological targets using single-walled carbon nanotubes (SWCNTs) [1]. Some attempts to design 3D layer-by-layer structures to improve the sensor sensitivity by increasing the density of the immobilized bioreceptors, were limited to the fabrication of enzyme sensors. Indeed, the enzyme sensors are used for the determination of small analyte such as glucose or catechol that can diffuse easily into multilayered enzyme 3D structures [2, 3]. In case of immunosensors, in addition to the bioreceptor unit (antigen) and analyte (antibody), an additional secondary antibody labelled with a redox enzyme has to be immobilized, which are several hundred nanometers in size. Therefore, optimal diffusion of all these biological receptors and the analyte remains a challenge for immunosensing applications. In this work, with the size of nanoparticles, the pore size of the scaffolds was optimized to immobilize a maximum amount of bioreceptors, and even enzymatic markers attached to large secondary antibodies.

EXPERIMENTAL DETAILS

The different 3D nanotubes/nanoparticles scaffolds were constructed on the platinum electrodes ($\Phi = 2$ mm). Nanotubes were purchased from Unidym, Sunnyvale, CA. Standard Carboxyl-Adembeads 0211 (100 nm) and masterbeads Carboxylic Acid 0215 (500 nm) were purchased from Ademtech. Magnetic microparticles (Dynabeads M-270 carboxylic acid (2.4 μm)) were purchased from Invitrogen.

For one and two level scaffolds, the first layer of nanotubes (20 μL, 0.1mg/mL in N-Methylpyrrolidine, NMP) was deposited by classical drop casting method [4]. Supplement levels were constructed by repeatedly deposition of nanoparticles (NPs) / microparticles (μPs) (5μL, 0.5% (v/v)) and nanotubes (10μL), followed by rinsing with water at each step. The constructed scaffolds were then functionalized by incubating in a solution of pyrene-β-cyclodextrin (2mM in NMP) by non-covalent interactions, which were then reinforced by electropolymerization and characterised at each step with scanning electron microscope using ULTRA 55 FESEM 176.

For glucose enzyme sensor, 5 μL of a phosphate buffer solution (PBS) containing GOX-B was deposited on the electrode surface via affinity interaction between biotin and and β-cyclodextrin [5], for 20 min at 4 °C. For immunosensor, the nanostructured electrodes were incubated with 5 μL of a PBS containing the biotinylated cholera toxin B subunit (antigen), anti-cholera toxin (anti-CT) antibody developed in a rabbit, biotin monoclonal anti-rabbit IgG (secondary antibody), and finally with cyclodextrin modified glucose oxidase, GOX-CD (enzyme marker) successively for 20 min each at 4°C. Both the electrodes were then rinsed with 0.1 M PBS two times. The concentration of all biological compunds used was 0.5 mg/mL.

The amperometric measurements were performed with a Tacussel PRG-DL potentiostat (Tacussel, France) controlled by the Echart software (eDAQ, Australia) in stirred phosphate buffer solution (PBS, 0.1 M, pH 7) at 25 °C. An Ag/AgCl electrode was used as a reference electrode and a Pt wire, served as counter electrode.

DISCUSSION

Scanning electron microscopy imaging

The morphologies of different nanoscaffolds were first characterized by scanning electron microscope as shown in figure 1.

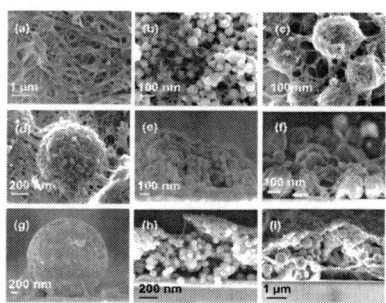

Figure 1. SEM images of (a) SWCNTs, (b) 100nm NPs, (c) 500 nm NPs, and (d) 2.4 μm μPs, covered with nanotubes. Cross-sections of one level scaffold made with100 nm, 500 nm NPs and μPs are shown in (e), (f) and (g), while (h) and (i) show the cross-sections for the two level scaffolds made with 100 nm and 500 nm NPs, respectively.

Figure 1(a) shows the formation of thin nanotube films on the surface. Figures 1(b) to 1(d) show the top views of the one level scaffolds for nanoparticles (100 nm, 500 nm) and microparticles, where nanotubes uniformly deposited over nanoparticles and microparticles leaving some vacancies can be clearly seen. Figures 1(e) to 1(i), represent cross-section views of one and two level scaffolds. For 100 nm nanoparticles, a more compact structure made of nanoparticles was observed, for the one as well as two level scaffolds, whereas for the 500 nm nanoparticles a homogeneous single layer of nanoparticle covered with nanotubes can be seen for one level scaffold and a more recognizable layer by layer structure with the two level scaffold. In case of microparticles, only few individual particles covered with nanotube were observed without any significant difference between the two scaffolds. It might be due to the heavy weight of microparticle, which just fell off during the washing procedure.

Amperometric measurements of glucose oxidase based enzyme sensor and immunosensor

GOX catalyzes the oxidation of glucose to gluconic acid by producing hydrogen peroxide out of oxygen regenerating the enzyme. H_2O_2 is then oxidized at the Pt-electrode at 0.6 V versus Ag/AgCl, leading to current increase, which is directly related to the glucose concentration. Calibration curves obtained by monitoring the amperometric response of various electrodes at varying glucose concentrations for glucose enzyme sensors are shown in figure 2. The response time of the glucose substrate was satisfyingly fast, ranging from 5 to 30 s.

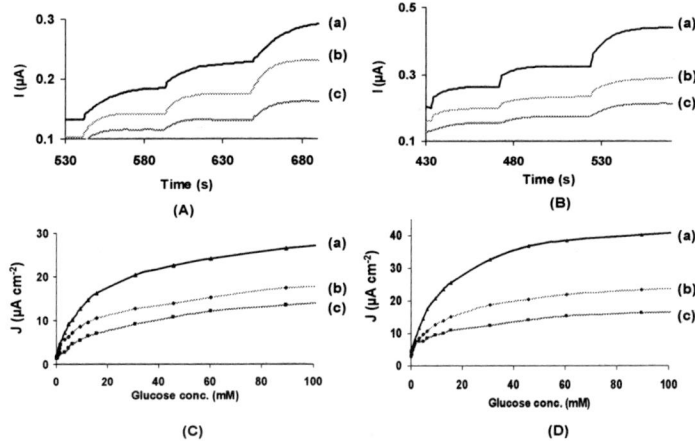

Figure 2. Steady state current – time responses of GOX-B enzyme sensors to successive injection of glucose, made with (A) one level and (B) two level scaffolds. Calibration curves for GOX-B enzyme sensor in presence of glucose, made with (C) one level and (D) two level scaffolds. Scaffolds made with (a) 100 nm NPs-SWCNTs, (b) 500 nm NPs-SWCNTs, and (c) 2.4µm µPs-SWCNTs.

The 100 nm nanoparticle/nanotube scaffolds, both the one and two level have the highest maximum current density (J_{max}) which corroborates to the maximum amount of enzyme that can be immobilized onto the surface. The values of J_{max} increased with the decrease in the size of particles, and increase in the scaffold levels. This indicates that the pore sizes created using 100 nm NPs allow unhindered diffusion of small glucose substrate and hence, thus produced H_2O_2 to the underlying electrode surface. The values of maximum current density along with the sensitivities obtained for different nanostructured scaffolds are listed in Table I.

Table I. Enzymatic performances of different nanoscaffolds for Glucose biosensor.

Scaffold Type	Nanotube layers	100 nm 1L	100 nm 2L	500 nm 1L	500 nm 2L	2.4μm 1L	2.4μm 2L
Max. current density J_{max} ($\mu A\ cm^{-2}$)	17.57	27.83	41.72	18.05	24.36	14.64	16.87
Sensitivity (mA $M^{-1}\ cm^{-2}$)	1.03	1.91	3.04	1.4	4.52	0.4	0.55

Where, 1L and 2L represents one and two level scaffolds respectively.

The electrode containing only nanotube layers, had the lowest current density and sensitivity as compared to other type of scaffolds, except for scaffolds made with microparticles where the values were remarkably comparable. This behaviour can be related to SEM observations, where few microparticles were present on the surface behaving more like nanotube layers.

Since the two level scaffolds have better performances than one level scaffolds, such two level scaffolds were evaluated for amperometric detection of anti-CT (500 μg/mL) for immunosensing applications. The detection of anti-CT was achieved using a sandwich immunoassay based on the immobilization of biotinylated cholera toxin B subunit, as the bio-recoginition layer onto the polypyrene β-cyclodextrin coated nanostructured electrodes by affinity interactions [5]. The analyte (anti-CT) was detected through enzymatic marker GOX-CD conjugated to biotinylated IgG antibodies (Figure 3).

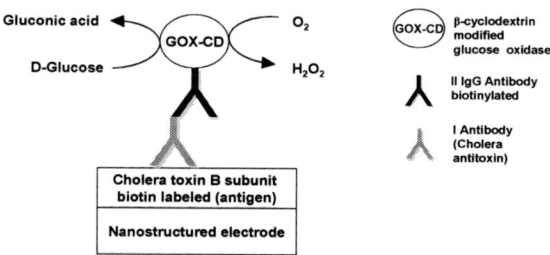

Figure 3. Amperometric immunosensor set up using polypyrene β-cyclodextrin coated nanostructured electrode and GOX-CD as enzymatic marker for the detection of cholera antitoxin (anti-CT).

The amperometric transduction of an immunoreaction consists in the post-labelling of the detected target (here, anti-CT) by an enzyme (here, GOX) able to catalyze the production or the consumption of electroactive species. The electrochemical oxidation or reduction of the latter at a constant potential applied to the immunosensor provides thus a current, which intensity is proportional to the amount of immobilized target. Figure 4 shows the histogram obtained by measuring the maximum current density of different scaffolds, which are related to the amount of immobilized antibody (anti-CT) in presence of glucose, along with the blind experiment performed for non-specific binding of GOX-CD labelled secondary antibody without antigen receptor on polypyrene β-CD modified SWCNTs scaffolds.

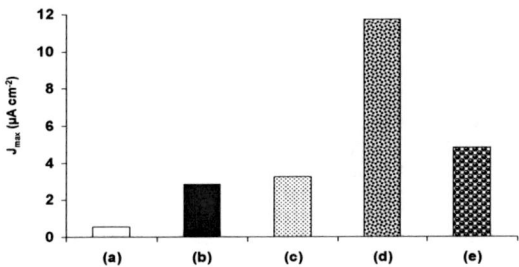

Figure 4. Histogram of the measured current density values for the GOX-CD labelled immunosensor for the detection of anti-CT in presence 0.3 M of glucose, obtained with different two level scaffolds. (a) Blind experiment for non-specific binding of GOX-CD labelled secondary antibody without antigen receptor, (b) SWCNTs scaffolds, (c) 100 nm NPs-SWCNTs, (d) 500 nm NPs-SWCNTs, and (e) 2.4 μm μPs-SWCNTs.

The change in maximum current density for non-specific binding of antibody represents an error of 5% compared to specific immune interactions. The comparison of the measured current densities of various scaffolds, clearly shows that the most sensitive amperometric immunosensor was based on the nanoscaffold constructed with 500 nm nanoparticle with higher max current density and sensitivity values (11.71 μA cm^{-2}, 116.2 μA M^{-1} cm^{-2}), as compared to microparticles (4.8 μA cm^{-2}, 29.7 μA M^{-1} cm^{-2}) and 100 nm (3.2 μA cm^{-2}, 47.44 μA M^{-1} cm^{-2}). Again, scaffold containing only nanotubes (2.8 μA cm^{-2}, 22.7 μA M^{-1} cm^{-2}), had lowest current density and sensitivity but comparable to that of microparticles.

It's evident that there is significant increase in the performance of immunosensor constructed with 500 nm NPs scaffolds, with optimal pore sizes to immobilize large number of bioreceptors which are several hundred nanometers in size and even, further allows unhindered diffusion of the H$_2$O$_2$ to the electrode surface, when compared to 100 nm NPs scaffold having more compact structure and comparatively small pore sizes causing hindrance in the permeation of bioprobes.

CONCLUSIONS

We proposed an innovative approach to immobilize a large amount of biological receptor molecules on such nanoscaffolds with an optimal surface/volume ratio keeping accessibility for the analyte. For the glucose enzyme sensor, the most appropriate nanoscaffolds were made with smaller nanoparticles i.e. 100nm. While in case of immunosensors, nanoscaffolds made with 500 nm nanoparticles performed best allowing permeation of secondary antibody further attached to any enzymatic marker. For both type of sensors, with the increase in nanoscaffolds level, the maximum current density and sensitivity increases. Therefore, these constructions will contribute as high sophisticated material for the development of biosensors or biofuel cells.

ACKNOWLEDGMENTS

The authors would like to thank the platform 'functionalization of surfaces and transduction' of the scientific structure 'Nanobio' for providing facilities The authors would also like to thank the French–Canadian Research Found (FCRF) for the Ph.D fellowship.

REFERENCES

1. R. J. Chen, H. C. Choi, S. Bangsaruntip, E. Yenilmez, X. Tang, Q. Wang, Y.-L. Chang, H. Dai, *Journal of the American Chemical Society*. **126**, 1563-1568 (2004).
2. M. Díaz-González, M. B. González-García, A. Costa-García, *Electroanalysis*. **17**, 1901-1918 (2005).
3. E. Katz, I. Willner, *Electroanalysis*. **15**, 913-947 (2003).
4. R. Haddad, S. Cosnier, A. Maaref, M. Holzinger, *Sensor Lett.* 7, 801–805 (2009).
5. M. Holzinger, M.Singh, and S.Cosnier, *Langmuir*.**28 (34)**, pp 12569–12574 (2012).

Mater. Res. Soc. Symp. Proc. Vol. 1700 © 2014 Materials Research Society
DOI: 10.1557/opl.2014.573

DNA-Assisted Dispersion of Multi-Walled CNTs in Epoxy Polymer Matrix

Susanna Laurenzi[1], Matteo Sirilli[1], Mirko Pinna[1] and M. Gabriella Santonicola[2,3]

[1]Department of Astronautic Electrical and Energy Engineering, Sapienza University of Rome,
Via Salaria 851-881, 00138 Rome, Italy
[2]Department of Chemical Materials Environmental Engineering, Sapienza University of Rome,
Via del Castro Laurenziano 7, 00161 Rome, Italy
[3]Materials Science and Technology of Polymers, MESA+ Institute for Nanotechnology,
University of Twente, 7500 AE Enschede, The Netherlands

Corresponding authors: susanna.laurenzi@uniroma1.it, mariagabriella.santonicola@uniroma1.it

ABSTRACT

The homogeneous dispersion of carbon nanotubes (CNTs) in a polymer matrix is a critical parameter that significantly affects the electrical and mechanical properties of CNT-based composite materials, and represents an important challenge to overcome during the manufacturing process of these materials. In our work we used double-stranded DNA to facilitate the dispersion of multi-walled CNTs in solution prior to the integration in epoxy resin PRIME 20 LV. Composites containing DNA-wrapped CNTs were prepared using sonication at 0.5 wt.% CNT loading and the dispersion level in the composite CNT/PRIME 20 LV was observed under an optical microscope. Nanoindentation experiments were conducted to determine the local mechanical properties of the CNT/PRIME 20 LV composites films after cure, showing a significant improvement in their distribution across the sample surface as a result of the enhanced CNT dispersion by DNA. An electrical test to assess the stability of the CNTs dispersion in the resin was developed by measuring the conductivity of the composite mixture before cure in time. Results of the electrical measurements indicate that the CNT/PRIME 20 LV mixture with DNA-wrapped CNTs is stable for several days after preparation.

INTRODUCTION

Nanocomposites based on carbon nanotubes are extensively investigated in different fields of research for their manifold exceptional properties. CNTs possess high strength, high modulus and very interesting electric and thermal features for a wide range of applications [1-5]. However, the performance of nanocomposites is strictly related to the CNTs dispersion in the polymer matrix. Enhanced dispersion of CNTs in a polymer matrix greatly improves the mechanical, thermal, electrochemical, optical and hydrophobic properties of the composite materials. The main problem that is encountered in reinforcing a polymer matrix with CNTs is to obtain a homogeneous dispersion thus avoiding that the CNTs turn into localized defects in the nano-reinforced composites. What makes CNTs dispersion extremely difficult is their large specific surface and large aspect ratio characteristic of the nanotubes [6-7]. In fact, the mutual attraction of the CNTs, due to van der Waals forces, drives their aggregation into clusters, even after mechanical disentanglement processes employed during fabrication. In order to overcome these problems, different approaches have been developed based on surface modification such as covalent and non-covalent functionalization methods [6-7].

The covalent functionalization of CNTs reduces their exceptional properties because chemical treatment breaks the CNT-structure introducing defects on its surface. At the same time, non-covalent functionalization mainly based on the use of dispersants, such as surfactants and polymers, are often not effective and the CNTs remain predominantly entangled. On the other hand, sonication of CNTs in DNA media (mostly single-stranded DNA) has gained increased attention due to the efficient nanotube dispersion that can be achieved [8]. The sonication process is necessary to provide an efficient DNA wrapping around the nanotubes with almost any sequence of single-stranded DNA and short double-stranded DNA, indicating that DNA binding to carbon nanotubes is very strong [9].

In this work, we present the effect of a recently designed dispersion approach, based on non-covalent functionalization of CNTs by DNA molecules, on the preparation of CNT/PRIME 20 LV composites films. Multiwalled CNTs were first functionalized with DNA, and then used to fabricate nanocomposites at CNTs loading of 0.5 wt. %. Different techniques were used to test the dispersion homogeneity of the composites. In particular, a setup for electrical measurements was developed to assess the stability of the CNT/PRIME 20 LV mixture in time.

EXPERIMENT

DNA-wrapped multiwalled CNTs (MWCNTs) were obtained by sonication of MWCNTs in DNA aqueous solution (1 mg/mL, DNA MW ~1300 kDa) for 60 min in cold bath to avoid DNA denaturation following a procedure reported previously [5]. The blend ratio between MWCNTs weight and volume of solution was 1:1 (mg/ml) since this proportion guarantees a stable mixture for days. After the sonication procedure, DNA molecules wrap along the nanotubes structure, acting as a highly efficient dispersing agent for the CNTs in aqueous solutions due to the negatively-charged phosphate groups on the DNA molecule [8]. The efficiency of the DNA-mediated dispersion was demonstrated to be independent of the length of the DNA molecules, considering that the duration of the sonication process might shorten the DNA strands [9, 10].

Before dispersion of the DNA-wrapped MWCNTs in the polymer matrix, the mixture aqueous solution was filtered in order to remove excess water from the DNA-wrapped MWCNTs, and then dried at 50 °C for several hours to remove any humidity residues. DNA-wrapped MWCNTs and pristine MWCNTs were dispersed at 0.5 wt. % in the epoxy matrix PRIME LV 20 (200 g) following the same two-step procedure: mechanical stirring for few minutes and sonication for about 1 h.

Nanocomposites were fabricated by adding the slow hardener at 26 wt. % with respect to the matrix content. Before starting the polymerization phase, the mixture was degassed to ensure the elimination of air bubbles formed during the stirring step. The specimens were realized by pouring the mixture in a silicon mold, and curing at 50 °C for 16 h. Specimens were fabricated at DNA-MWCNT loading of 0.5 wt. %.

Electrical tests on the uncured mixture were performed using the setup showed in Figure 1: two electrodes were immersed in the blend and the electrical resistance varying the alternating current in the range of 1-10 MHz was measured using an Agilent LCR meter. The measurements were performed on both equivalent mixtures of pristine MWCNTs and DNA-wrapped MWCNTs.

Nanoindentation tests were performed on cured specimens with a NanoTest instrument equipped with a Berkovich-type indenter tip. The indenter had a load resolution of 150 nN and a depth resolution of 0.1 nm. For each specimen 16 points randomly located at the surface were analyzed, and for each point the test was repeated six times to have a statistically significant result. Particular attention was given to include sample points on the vertical face of the specimen across its thickness: if the MWCNTs are not well dispersed and the dispersion is not stable, a material gradient occurs through the thickness due to the precipitation of the carbon nanoparticles inside the mold cavity.

Optical micrographs for analysis of dispersion of the CNT/PRIME 20 LV mixtures on the microscale were taken on small volumes casted on microscopy slides.

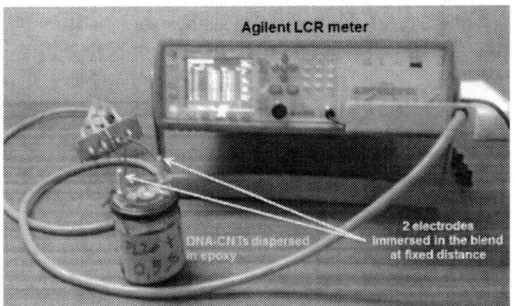

Figure 1. Setup for electrical measurements of uncured DNA-MWCNTs/epoxy resin mixtures showing the different components of the system.

DISCUSSION

Optical microscopy

In the optical micrographs in Figure 2 we compare the different dispersion level obtained with 0.5 wt. % of pristine MWCNTs (Figure 2a) and 0.5 wt. % of DNA-functionalized MWCNTs (Figure 2b) in the epoxy resin PRIME LV 20.

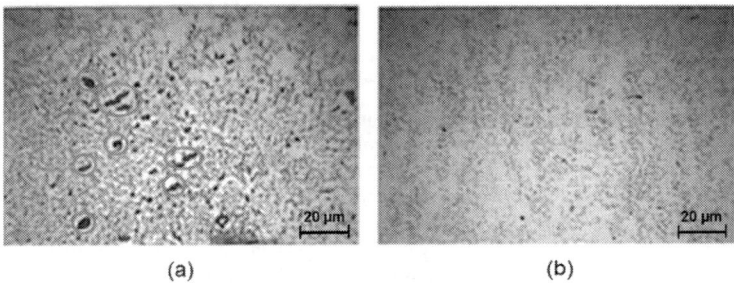

(a)   (b)

Figure 2. Optical micrographs at 40X magnification of MWCNTs (0.5 wt. %) dispersed in resin PRIME LV 20: (a) dispersion of pristine MWCNTs and (b) dispersion DNA-wrapped MWCNTs. Larger MWCNTs aggregates are indicated by red circles.

As it can be observed, even if the two types of mixture underwent the same dispersion procedure, the pristine MWCNTs form large agglomerates in the matrix (Figure 2a), whereas the DNA-functionalized MWCNTs show a quite uniform distribution with sporadic small clusters.

Nanoindentation

The nanoindentation tests were performed to determine the hardness and the elastic modulus of the nanocomposites in different points of the surface. This method was adopted to evaluate the dispersion of the nanofillers: the smaller is the standard deviation of the measurements across the surface and more homogenous is the material, and therefore a good dispersion of the MWCNTs in the epoxy resin was obtained. Table I summarizes the results of the indentation tests performed on cured specimens of neat resin, resin reinforced with pristine MWCNTs, and resin reinforced with DNA-wrapped MWCNTs.

Table I. Hardness and reduced modulus from nanoindentation tests for specimens fabricated with neat resin, pristine-MWCNTs reinforced resin and DNA-MWCNTs reinforced resin.

Material	Hardness (MPa)	Red. Modulus (GPa)
Neat resin	0.24 ± 0.002	4.50 ± 0.013
pristine-MWCNTs reinforced epoxy	0.25 ± 0.150	4.70 ± 0.199
DNA-MWCNTs reinforced epoxy	0.29 ± 0.030	4.50 ± 0.008

By comparing the data in Table I, it can be observed that the nanocomposites fabricated with pristine MWCNTs present large scattering of the measurements, which indicates that the material is not homogenous. As a consequence, the improvement of the hardness with respect to the neat resin is negligible. On the other hand, for DNA-MWCNTs reinforced epoxy samples the surface hardness increases with respect to the pristine-MWCNTs sample by 16%, and the standard deviation is significantly lower (80%). These results suggest that the enhanced dispersion of the MWCNTs obtained after non-covalent wrapping by DNA improves the surface properties of the nanocomposite specimens after curing.

Electrical measurements

The electrical measurements were performed to evaluate the stability of the uncured CNTs/epoxy resin mixture in time. Figure 3 shows the trend of electrical resistance as a function of current frequency at different times for the blends of pristine-MWCNTs reinforced epoxy (Figure 3a) and DNA-MWCNTs reinforced epoxy (Figure 3b). The electrical resistance of the pristine-MWCNTs reinforced epoxy is four orders of magnitude less than for DNA-MWCNTs reinforced epoxy resin. This result can be explained considering the electrical properties of MWCNTs and DNA molecules. While carbon nanotubes are well known to be conductive nanofillers that enhance the electrical conductivity of an insulating epoxy matrix [11], DNA molecules are strongly dielectric [12]. Therefore the hybrid DNA-MWCNTs nanostructures become insulated after DNA wrapping. This change is reflected in the large difference in the electrical resistance using pristine MWCNTs or DNA-MWCNTs nanofillers. In this work, we focused on the stability in time of the dispersions. The solution of pristine-MWCNTs reinforced epoxy appears quite unstable in time: the pristine MWCNTs tend to aggregate into bundles because of van der Waals forces. The entanglements fluctuate inside the polymer with weak interactions and, after a certain time, they tend to settle in the sample container thickening around the electrodes. This phenomenon explains the trend observed after 90 h, which is also confirmed by visual inspection. On the other hand, DNA-MWCNTs reinforced epoxy show a more constant value of the electrical resistance at different times (Figure 3b), and the dispersion appears stable even after several days.

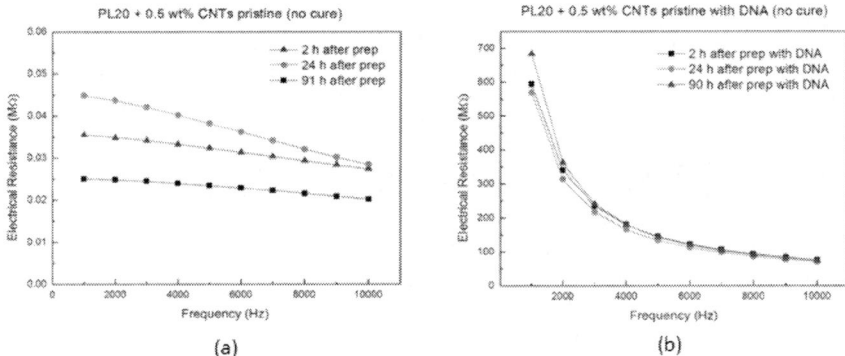

(a)   (b)

Figure 3. Electrical resistance as function of current frequency at different times for the pristine MWCNTs reinforced epoxy (a) and DNA-MWCNTs reinforced epoxy (b).

CONCLUSIONS

The outstanding properties of carbon nanotubes drive towards many potential applications in polymeric composite materials. Their use is still limited by technological challenges, mainly related to the integration of CNTs into the polymeric phase. To obtain a high dispersion level of CNTs avoiding the formation of large clusters is of fundamental importance for practical applications and manufacturing of composite materials. In the latter case, the carbon nanotubes act as defects inside the composites reducing its mechanical performance.

In this work, we have showed a method to disperse multiwalled carbon nanotubes into an epoxy matrix, adopting a non-covalent surface modification of the nanotubes with DNA molecules. The dispersion level obtained with DNA-functionalized MWCNTs in the resin PRIME LV 20 was investigated by optical, electrical and nanoindentation tests and results were compared with those obtained using pristine MWCNTs. The findings reported here indicate that DNA-wrapping of MWCNTs is an efficient way to obtain a homogenous and stable dispersion of CNTs in an epoxy resin matrix for several days, thus opening the way to novel processing routes for these composite materials.

ACKNOWLEDGMENTS

The authors would like to thank Dr. Antonia Simone (Thales Alenia Space) for the nano-indentation tests. M.G.S. acknowledges financial support from the Italian Ministry of Education, University and Research (MIUR) through the Montalcini Program.

REFERENCES

1. S. Laurenzi, R. Pastore, G. Giannini, M. Marchetti, *Compos. Struct.* **99**, 62 (2013).
2. S. S. Rahatekar, M. Zammarano, S. Matko, K. K. Koziol, A. H. Windle, M. Nyden et al. *Polym. Degrad. Stab.* **95**, 870 (2010).
3. Z. Fan, M. H. Santare, S. G. Advani, *Compos. Part A* **39**, 540 (2008).
4. M. G. Santonicola, S. Laurenzi, M. Marchetti, *Proceedings of the 61st International Astronautical Conference 2010,* **13**, 10938-10940 (2010).
5. S. Wang, R. Liang, B. Wang, C. Zhang, *Carbon* **47**, 53 (2009).
6. P.-C. Ma, N. A. Siddiqui, G. Marom, J.-K. Kim, *Compos. Part A* **41**, 1345 (2010).
7. L. Vaisman, H. D. Wagner, G. Marom, *Adv. Colloid Interface Sci.* **128**, 37 (2006).
8. M. Zheng, A. Jagota, E. D. Semke, B. A. Diner, R. S. McLean, S. R. Lustig, R. E. Richardson, N. G. Tassi, *Nat. Mater.* **2**, 338 (2003).
9. X. Tu and M. Zheng, *Nano Res.* **1**, 185 (2008).
10. T. L. Mann, U. J. Krull, *Biosens. Bioelectron.* **20**, 945 (2004).
11. J. Sandler, M. S. P. Shaffer, T. Prasse, W. Bauhofer, K. Schulte, A. H. Windl, *Polymer* **40**, 5967 (1999).
12. A. J. Storm, J. van Noort, S. de Vries, C. Dekker, *Appl. Phys. Lett.* **79**, 3881 (2001)

Mater. Res. Soc. Symp. Proc. Vol. 1700 © 2014 Materials Research Society
DOI: 10.1557/opl.2014.605

Methods for Dispersion of Carbon Nanotubes in Water and Common Solvents.

Boris I. Kharisov, Oxana V. Kharissova, Ubaldo Ortiz Méndez
Universidad Autónoma de Nuevo León, Monterrey 66450, Mexico.

ABSTRACT

Contemporary methods for dispersion of carbon nanotubes in water and non-aqueous media are discussed. Main attention is paid to ultrasonic, plasma techniques and other physical techniques, as well as to the use of surfactants, functionalizing and debundling agents of distinct nature (elemental substances, metal and organic salts, mineral and organic acids, oxides, inorganic and organic peroxides, organic sulfonates, polymers, dyes, natural products, biomolecules, and coordination compounds).

INTRODUCTION

The carbon nanotubes (CNTs), classic objects in nanotechnology, form bundle-like structures with very complex morphologies with a high number of Van der Waals interactions, causing extremely poor solubility in water or organic solvents. It is difficult to prepare stable aqueous dispersion of CNTs; their insolubility has been a limitation for the practical application of this unique material. Proper dispersion of CNTs materials is important to retaining the electronic properties of nanotubes. The redissoluble functional compound/CNTs composites are needed for post processing because CNTs dispersions usually easy aggregate and therefore make additional processing very difficult. Nonfunctionalized CNTs can be solubilized in suitably chosen organic solvents and, furthermore, their solubility could be understood in terms of the Hansen Solubility Parameters (HSPs). As it will be shown below (Table 1 and throughout the text), various dispersion approaches are based on physical (dielectrophoresis, gel electrophoresis, density gradient ultracentrifugation, plasma and irradiation tecniques, and chromatography) and chemical (diazonium salts, ozonolysis, functionalization with porphyrins, bromine, amines, pyrene, DNA, peptides, polymers, etc.) methods.

Table 1. Overview of physical and chemical methods for CNTs dispersion.

Technique	Observations
Ultrasonication	One of the most frequently used methods. It can be used with or without use of surfactants, simultaneous chemical treatment or dielectrophoresis.
Plasma method	RF plasma discharge in O_2 or $Ar/O_2/C_2H_6$.
Irradiation	Mainly γ-irradiation.
Ozonolysis	Generaly by UV/ozone treatment.
Functionalization	This is carried out with use of: a) inorganic compounds, such as carbon allotropes (nanodiamonds, graphene oxide), iodine doping, mineral acids and their mixtures, CO_2 and SiO_2, peroxides, inorganic salts; b) organic compounds: diazonium salts, organic acids and salts, sulfonates, amines, porphyrins, pyrene and other polyaromatic compounds, DNA, biomolecules, polymers, natural products, etc.

The main objective of this article is to generalize briefly main recent (2009-2013) methods for CNTs dispersion in distinct media and to represent contemporary trends in this field.

PHYSICAL METHODS

Ultrasonication is often used to disperse nanoparticles in aqueous solution. However, a good dispersion of nanoparticles in aqueous solution is not always achieved, due to the fact that incoming ultrasonic waves in liquid are usually reflected and damped at the gas/liquid interface. For the case of carbon nanotubes, this is a classic debundling method: the MWCNTs can be effectively ultrasonically dispersed in the water solution[1] and a series of organic solvents.[2] A considerable number of reports, dedicated to the use of surfactants and functionalizing agents for CNTs dispersion/solubilization, imply simultaneous application of ultrasound of distinct power, weak or elevated. A union of both ultrasonic sources is also known.

Plasma and irradiation techniques. Distinct types of plasma techniques have been applied to improve CNTs dispersibility, frequently without surfactants, giving different quantitative results. Thus, carbon nanotube microcapsules were prepared[3] by oil in water (O/W) Pickering emulsions without any surfactant used. CNTs were treated with oxygen plasma (radio frequency of 13.56 MHz) at a power of 100 W and a pressure of 200 mTorr for several different periods. The oxygen plasma treatment introduced several hydrophilic groups on carbon nanotubes resulting in the improved aqueous dispersion. Radiation methods have been applied for CNTs dispersion too. Thus, a method for highly efficient functionalization of SWCNTs by DNA wrapping included exposure of SWCNTs to γ-irradiation (50 kGy), which lowered by one order of magnitude the amount of single stranded deoxyribonucleic acid (ssDNA) required for SWCNT modification.[4] While γ-irradiation in three different media significantly improved the process of SWCNT dispersion, irradiation in ammonia was the most efficient. The γ-irradiated SWCNTs functionalized with ssDNA were stabilized by electrostatic forces. The authors suggested that γ-irradiation can significantly improve the functionalization of SWCNTs with DNA.

FUNCTIONALIZATION OF CNTs THROUGH CHEMICAL METHODS

Inorganic compounds. Among them we note *other carbon allotropes* as nanodiamonds (NDs) and graphene oxide, which were applied for improvement of CNTs dispersion. Thus, nanodiamond particles were used[5] to disperse CNTs, allowing the formation of their stable colloidal suspensions. As an example of other *compounds in elemental form*, chemical functionalization of MWCNTs was carried out[6] by UV/ozone treatment. The presence of oxygen-containing groups (such as carboxylic, quinine, and hydroxyl groups) on the MWCNT surfaces by UV/ozone treatment was confirmed resulting in dispersion stability better than for pristine MWCNTs in polar solvents. Iodine-doping into SWCNTs can be effectively done using an electrochemical method[7] and can be easily and finely controlled by changing the polarity. But, in our opinion, the most intriguing example in this section is an unusual application of $Na-NH_{3(liq.)}$ system, well-known in classic inorganic chemistry courses. Thus, an intrinsically scalable method for SWNT dispersion and separation, using reductive treatment in sodium metal-ammonia solutions was discussed,[8] optionally followed by selective dissolution in a polar aprotic organic solvent. SWNTs can indeed be unbundled to give individual tubes in solution by reductive charging in ammonia. Other inorganic compounds used for CNTs dispersion were _CO_$_2$ and _SiO_$_2$ (the dispersion of MWCNTs at the very low milligram level was achieved by effervescence due to the *in situ* generation of carbon dioxide like in some traditional pharmaceuticals), _inorganic acids and their mixtures_,[9] _inorganic salts_ (NaCl/NaI[10]), and _inorganic peroxides_ (persulfates[11]).

Organic acids and salts are more widespread, for instance carminic acid or benzenetricarboxylic acid (BTC), sodium dodecylbenzene sulfonate (SDBS) and sodium dodecyl sulfate (SDS), which are very popular and mostly frequently used due to their outstanding properties for CNTs dispersion, first of all dispersion stability in distinct temperature conditions and solvents.[12] Other sulfonate-containing surfactants are rare; among them, we note water-soluble polyaniline blend poly(sodium 4-styrenesulfonate), (PANI·PSS), which was used[13] to disperse MWCNTs by noncovalent surface modification. Several other organic salts, applied for CNTs dispersion, are also known, for instance dodecyl quaternary ammonium bromides. As an example, dodecyl trimethylammonium bromide (DTAB) and sodium octanoate (SOCT) were found to form exceptionally stable MWNT dispersions.

Other organic and coordination compounds. Other organic non-biological substances discussed below are mostly aromatic and polyaromatic compounds, frequently containing heterocycles, for instance porphyrin (and its metal derivatives) or thiophene. As an example of an S-heterocycle, the hydrogenation of 2-nitrothiophene gave 2-aminothiophene that was used[14] for amidation of SWCNTs functionalized with carboxylic acid groups (SWCNT-COOH). In these modified carbon nanotubes (SWCNT-CONHTh), the thiophenes were covalently attached to the SWCNTs *via* amide linkages. The modified SWCNTs showed enhanced solubility, and thus better dispersion in common organic solvents and were used as dopant in polymer-fullerene photovoltaic cells. In case of amines (RNH$_2$ = 4-aminopyridine, 5-aminoisophthalic acid, and *p*-anisidine, Fig. 1), an enhanced solubilities of functionalized SWCNTs were reached; for example, in the last case the solubility in common solvents was as follows (mg/m L), acetone (0.076), water (0.084), DMF (0.082), ethanol (0.078), toluene (0.09) and ethyleneglycol (0.09).[15] Experimental conditions of the synthesis of functionalized SWCNTs with a series of similar ligands are presented in Table 2 and the solubility data of the functionalized SWCNTs – in Fig. 2.

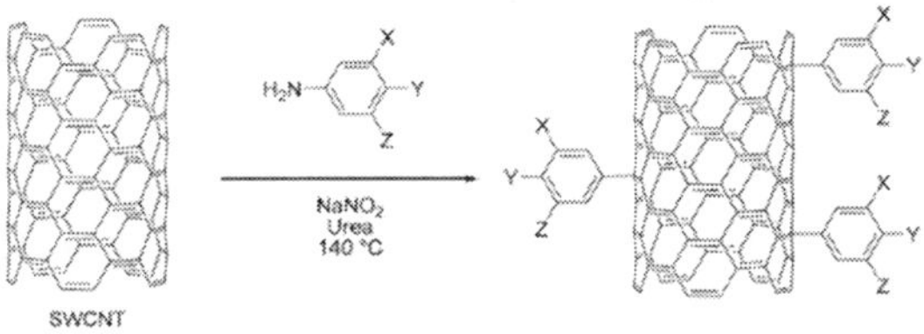

Figure 1. Synthesis of functionalized carbon nanotubes with aromatic amine derivatives.

Table 2. Functionalization of SWCNTs with amine ligands in molten urea (140°C).

SWCNTs, mg	Compound	Its weight, g	NaNO$_2$, g	Urea, g
50.0	4-Anisidine	1.58	1.16	60
50.0	4-Aminopyridine	2.88	1.16	60
50.0	2-Aminothiazol	2.06	1.16	60
50.0	4- Sulfanilamide	1.68	1.16	60

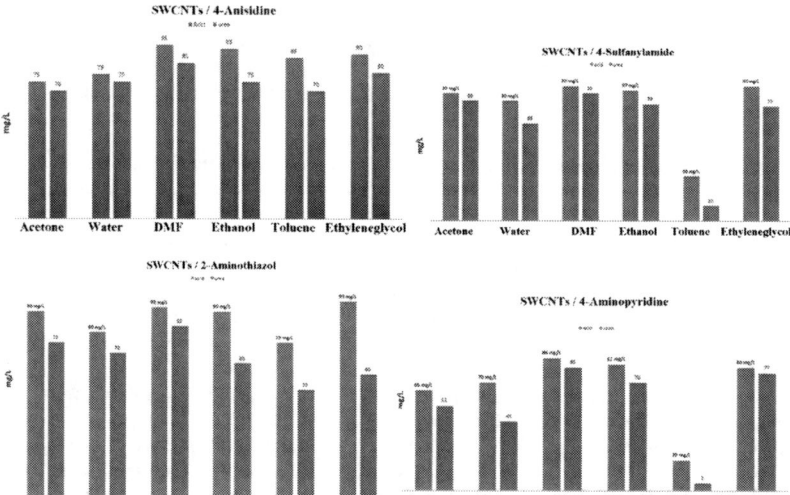

Figure 2. Solubility of amino-functionalized SWCNTs (AFSWCNTs) in solvents: blue color – obtained by classic method in HNO$_3$ medium (80°C), red color –obtained in molten urea.

Polymer-assisted dispersion of CNTs. Currently, the dispersion processes of CNTs in polymers or assisted by presence of polymers are the object of serious attention.[16] In case, if a polymer can be solubilized in solvents, its CNTs composites could also be dispersed; obviously, this is not an only condition for the possibility of polymer-assisted CNTs dispersion. In case of most simple polymers, such as poly(vinyl alcohol) (PVA), a low-temperature synthesis method of amorphous carbon nanotubes (a-CNTs) and PVA composite thin films were reported.[17] Different polymers can be used by stepwise preparation of composites. Thus, PEGylated MWNTs were prepared for the successive fabrication of poly(vinyl alcohol) PVA/MWNTs nanocomposite film by solution casting.[18]

The surface modified MWNTs showed a good colloidal stability in water and improved dispersion stability in aqueous PVA solution.

Other dispersing agents are *natural products* {for example, the gum arabic (GA)[19]}, *sugars* and *biomolecules,*[20] among others. Special comparative studies and techniques for understanding the CNTs dispersion included *dispersion of CNTs in solvents,*[21] *comparison of surfactant abilities,*[22] use of *combinations of surfactants,*[23] as well as *UV- and visible light influence,*[24] among others.[25]

CONCLUSIONS

A series of contemporary techniques are being used for CNTs solubilization, from physical (classic ultrasound or plasma treatment) to chemical and biological, applying inorganic (other carbon allotropes, iodine, metallic sodium in liquid ammonia, CO_2, peroxides, metal salts and mineral acids) and organic (acids, salts, polymers, dyes, natural products and biomolecules) compounds, as well as some metal complexes. It has been suggested that van der Waals interaction, π-π stacking interactions between aromatic rings in organic compounds and CNTs, and hydrophobic interaction are major factors that are responsible for the CNTs dispersion. Choosing surfactants, to be able to stabilize carbon nanotubes in water, it is necessary to employ such dispersing agents that a) strongly adsorb on the nanotube surface, b) present hydrophilic groups, better if rigid, that extend toward the aqueous phase, c) are not very mobile on the nanotube surface, and d) show aggregates with structure dependent on nanotube diameter and chirality.

REFERENCES

[1] L. Cui. *Adv. Mater. Res.*, **641-642** (1), 436-439 (2013).

[2] J.L. Bahr, E.T. Mickelson, M.J. Bronikowski, R.E. Smalley, J.M. Tour. Chem. Commun., 193–194 (2001).
[3] W. Chen, X. Liu, Y. Liu, H.-I. Kim. Materials Letters, **64** (23), 2589-2592 (2010).

[4] S.P. Jovanović, Z.M. Marković, D.N. Kleut, N.Z. Romević, V.S. Trajković, M.D. Dramićanin, B.M. Todorović Marković. Nanotechnology, **20** (44), art. no. 445602 (2009).
[5] S. Ciftan Hens, G. Cunningham, G. McGuire, O. Shenderova. Nanoscience and Nanotechnology Letters, **3** (1), 75-82 (2011).

[6] S. Kim, Y.-I. Lee, D.-H. Kim, K.-J. Lee, B.-S. Kim, M. Hussain, Y.-H. Choa. Carbon, **51** (1), 346-354 (2013).
[7] H. Song, Y. Ishii, A. Al-Zubaidi, T. Sakai, S. Kawasaki. Physical Chemistry Chemical Physics, **15** (16), 5767-5770 (2013).

[8] S. Fogden, C.A. Howard, R.K. Heenan, N.T. Skipper, M.S.P. Shaffer. ACS Nano, **6** (1), 54-62 (2012).
[9] J. Singh, M.C. Kothiyal, D. Pathania. International Journal of Theoretical and Applied Science, **3** (2), 15-20 (2011).

[10] S. Ou, S. Patel, B.A. Bauer. Free energetics of carbon nanotube association in pure and aqueous ionic solutions. Journal of Physical Chemistry B, **116** (28), 8154-8168 (2012).

[11] L. Zhang, Y. Hashimoto, T. Taishi, Q.-Q. Ni. Applied Surface Science, **257** (6), 1845-1849 (2011).
[12] A.Y. Vlasov, A.V. Venediktova, D.A. Videnichev, I.M. Kislyakov, E.D. Obraztsova, E.P. Sokolova. Physica Status Solidi (B) Basic Research, **249** (12), 2341-2344 (2012).

[13] E. Detsri, S.T. Dubas. Applied Mechanics and Materials, **229-231**, 223-227 (2012).

[14] M.M. Stylianakis, J.A. Mikroyannidis, E. Kymakis. Solar Energy Materials and Solar Cells, **94** (2), 267-274 (2010).
[15] O. Lucas Flores, O.V. Kharissova, U. Ortiz Mendez, H. Leija Gutiérrez, E. de Casas Ortiz, B.I. Kharisov. Journal of Chemistry, Article ID 573570, 8 pages (2013).
[16] Hang Woo Lee et al. Small, **5** (9), 1019–1024 (2009).
[17] D. Banerjee, A. Jha, K.K. Chattopadhyay. Macromolecular Research, **20** (10), 1021-1028 (2012).

[18] M.J. Kim, J. Lee, D. Jung, S.E. Shim. Journal of Macromolecular Science, Part A: Pure and Applied Chemistry, **47** (6), 588-594 (2010).

[19] B. Wang, Y. Han, K. Song, T. Zhang. Journal of Nanoscience and Nanotechnology, **12** (6), 4664-4669 (2012).

[20] H. Leinonen, M. Pettersson, M. Lajunen. Carbon, **49** (4), 1299-1304 (2011).

[21] H. Li, J.C. Nie, S. Kunsági-Máte. Chemical Physics Letters, **492** (4-6), 258-262 (2010).
[22] P. Kumar, H.B. Bohidar. Colloids and Surfaces A: Physicochemical and Engineering Aspects, **361** (1-3), 13-24 (2010).

[23] V. Datsyuk, P. Landois, J. Fitremann, A. Peigney, A.M. Galibert, B. Soulaa, E. Flahaut. J. Mater. Chem., **19**, 2729-2736 (2009).
[24] C.-Y. Chen, C.T. Jafvert. Environmental Science and Technology, **44** (17), 6674-6679 (2010).
[25] J.M. Hughes, D. Aherne, S.D. Bergin, A. Oneill, P.V. Streich, J.P. Hamilton, J.N. Coleman. Nanotechnology, **23** (26), art. no. 265604 (2012).

AUTHOR INDEX

SUBJECT INDEX

adhesion, 37

bonding, 9

C, 69, 85, 109
chemical vapor deposition (CVD)
    (chemical reaction), 3
colloid, 91
composite, 37, 55, 103

defects, 9, 29
devices, 17
dispersant, 103, 109

electrical properties, 17
electron irradiation, 29
electronic material, 91

fiber, 37

nanoscale, 47, 69, 79

nanostructure, 3, 29, 55, 61, 97, 103
nucleation & growth, 3

optical, 79

polymer, 61, 109
purification, 85

Raman spectroscopy, 9, 69

scanning electron microscopy
    (SEM), 55, 97
scanning probe microscopy (SPM),
    47
self-assembly, 47
semiconducting, 85
simulation, 17, 61, 79
solution deposition, 91
surface chemistry, 97

CPSIA information can be obtained at www.ICGtesting.com
Printed in the USA
LVOW10*1253241014

410356LV00002B/2/P